Added Value in Design and Construction

Added Value in Design and Construction

Allan Ashworth & Keith Hogg

Pearson Education Limited
Edinburgh Gate, Harlow
Essex CM20 2JE, England
and Associated Companies throughout the world

First published 2000

British Library Cataloguing in Publication Data
A catalogue entry for this title is available from the British Library

ISBN 0-582-36911-8

Set by 35 in Times 10/12pt and Helvetica
Printed in Singapore

Contents

Foreword

The service provided by the construction industry, sadly and not without some justification, has in recent years been under attack for failing to provide clients with what they really require from their buildings: that they are built to the appropriate quality, on time and within budget. If we collectively acknowledge that there is something wrong with our industry, it is important that we seek to remedy this situation through research and development, and the application of appropriate new skills and techniques. The clients of the construction industry require and deserve a first-class service that is both relevant and responsive to their needs. Therefore, in order to meet these challenges in a world of continuing and accelerating change, it is vital that we continually acquire new knowledge, develop new skills and apply these skills in practice. Those of us involved in the process of designing and constructing buildings must recognise this need, as much as any other manufacturer or service provider trying to survive in the global economy.

Continuing concerns about the industry have led to several important and influential publications which have highlighted the problems that we face, and have outlined a suggested framework of improvements. Perhaps one of the most significant of these, the Latham Report published in 1994, argued, in my view convincingly, the need for change at all levels in the industry. Despite this and other prominent recognition and promotion of new methods and approaches to managing the industry, there appears to be some resistance to the acceptance and utilisation of new ideas and methodologies in practice. Evidence suggests that action and results are not an automatic follow-on from proclamation.

Allan Ashworth and Keith Hogg have written a book that examines in detail a wide range of techniques and strategies that can be applied during the design and construction process, and makes clear the argument for the use of each in varying situations. It reflects the changing needs of an industry that is to some extent constrained by its own unique tradition, and is seemingly bound by confused values and hierarchical relationships that often work against the best interests of clients. The book is comprehensive in its coverage of practices and techniques. Some of these, for example whole life costing, have been used for many years but now require rejuvenation. Others recently established are regarded by most within the industry as being still in their infancy, and are not correctly used, if at all.

Clients, above all, are concerned with getting value for money and require our help in achieving this. The ethos of the book is clearly relevant to the needs of the industry that must adapt for the new millennium and it is therefore timely in its publication. I am pleased to add my support and believe that this book provides both guidance and encouragement to all of us who seek to add value in the normal course of our work. Without this mission, there seems little point to our professional existence.

Peter Roe FRICS FFB, January 2000
Past President of the Quantity Surveyor's Division of the Royal Institution of Chartered Surveyors. Partner A. E. Thornton-Firkin and Partners Chartered Surveyors and Construction Consultants.

Preface

It is not cheaper things that we want to possess, but expensive things that cost less.

(John Ruskin)

A family house at the beginning of the twentieth century cost approximately the same as a family car. By the beginning of the twenty-first century the ratio between the two was approximately 5:1.

This book describes principles and techniques that can be used for adding value to the process of constructing buildings. It emphasises the shift in practices that have occurred during the latter part of the twentieth century. This shift that has raised the importance of doing more with less – a feature that is now commonplace in all of our lives in all kinds of activity.

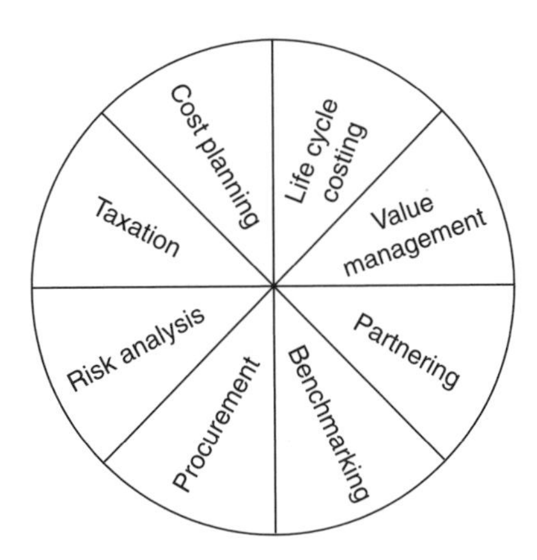

The techniques have been selected from a wide range of possible techniques that are available for use in the construction industry. Some of these techniques are tried and tested and have been enhanced since they were first introduced. Others are more recent, in some cases the principles have been borrowed from other industries. The techniques have been chosen since, among the array that are available, these perhaps offer the best potential in attempting to more closely match a client's objectives and desires with reality. They have also been partly introduced in response to criticism from inside and outside the industry.

Reducing building costs, or adding value, is beneficial not only for clients of the construction industry but also for society as a whole. If more hospitals, schools and houses can be built for the same capital outlay, then the benefits for the Government's financial strategy are obvious. Adding value offers public and private sector clients of the construction industry the ultimate reward. Reducing building costs by 30% (the figure suggested by the construction industry) was initially greeted with horror and disbelief. The fact that it has already been achieved on some projects in the United Kingdom, areas of the oil industry and by some firms overseas, most notably in the USA, suggests that it is now more likely to go some way towards becoming a reality. It should also be noted that this is not a new initiative unique only to the UK.

This practice of adding value is not a phenomenon of the construction industry in the United Kingdom. Like many good philosophies the principles are adopted world-wide. Should the UK construction industry choose to ignore the principles involved, then it will be left behind in a world market. The principles involved do not focus on the top tier of industry but among all firms, both small and great.

Allan Ashworth & Keith Hogg

About the authors

Allan Ashworth MSc ARICS is a former Courses Director at the University of Salford and HMI (Her Majesty's Inspector) in the Department of Education for construction education. He is currently a visiting professor in construction education at Liverpool John Moores University, England, and the adjunct professor in quantity surveying at UNITEC in New Zealand. He is a consultant with the Quality Assurance Agency for Higher Education (UK) and with the Royal Institution of Chartered Surveyors. He is the author of several standard textbooks.

Keith Hogg BSc ARICS PGCE is a senior lecturer in the Department of Surveying at The Nottingham Trent University where, until recently, he was the course leader for the BSc (Hons) in Quantity Surveying. He is now responsible for external consultancy within the department and is active in several areas of research. He was formerly employed as a quantity surveyor in private practice in the UK and overseas.

Chapter 1

The nature of added value

1.1 Introduction

The original and main use of buildings was for the purpose of shelter. However, today this use has been extended to many other purposes associated with production and investment, work and leisure. Buildings are no longer constructed as in the distant or recent past, nor are they financed and paid for in the same way. As design and construction have become more complex, the economics associated with these activities have also become more comprehensive. It is no longer sufficient to think of buildings in terms of their capital building costs alone, but to measure costs through a whole cost or holistic approach. This does not just envisage the buildings themselves but the land and plant costs and the way in which the buildings or projects are to be used. Today the application of creative building economics is as important as creative design and creative construction. The costs can therefore no longer be left to chance, wishful thinking or an act of God. Many building owners (Egan, 1998) have an enthusiastic desire to reduce the overall costs or to increase the long-term benefits achieved or, more preferably, to secure a combination of both of these objectives.

1.2 Value

Value is of all things a very subjective judgement. It is somewhat like beauty, where its assessment is largely in the eye of the beholder. The beholder, as far as building and construction projects are concerned, is the client who commissioned the project. Value constitutes a measure between supply and demand. An increase in the value of an object is achieved either through an increase in its demand or a decrease in its supply. Value also represents the usefulness and scarcity of an object relative to other objects or commodities. Value, therefore, gives an indication of both scarcity and usefulness when compared with other items. Value is regarded as a complex entity made up of scarcity, utility, costs of production, worth in use, value in exchange and marginal utility. It is influenced by the conditions of supply and demand.

Raftery (1991) points out that, in theory, it is expected that more will always be paid for a useful commodity than for one that is less useful. However, this might not always be the case and there are many examples that can be used to illustrate this point. The paradox of value illustrated by comparing water and diamonds suggests that the former has a low value but high utility whereas the diamond has a high value but low utility.

The economic theory of marginal utility, where the value of an additional unit is often related to the proportion of units already owned, suggests that value is only a relative term. Value is not intrinsic within an item but it is the relationship placed by someone on a particular item. Different individuals will therefore choose to value different items in different ways. Burt (1975) defines maximum value as a level of quality from a least cost, the highest level of quality for a given cost or from an optimum compromise between the two.

Kelly and Male (1993) suggest that value can be measured in currency, effort, exchange or on a comparative scale, that reflects the desire to obtain or retain an item, service or ideal. Authors often prefix value with different terms such as use, exchange, esteem, aesthetic, judicial, moral, etc., in order to highlight different value situations that reflect different value judgements. Use value, for example, is a measure of the function of the item. A chisel, for example, has a high value as a tool for cutting wood but a low value if it is used as a screwdriver, although it could be used for both tasks. Exchange value is the amount for which an item may be sold. Esteem value is the amount an owner or user is prepared to pay for prestige or appearance.

Value from a client's perspective is the relationship between quality and price (Morton, 1987). Quality can be defined as the degree of excellence of a product, trait, characteristic or attitude. These definitions can be consolidated and the term presented as the relationship between cost and the reliability or performance of a product. The value must be capable of measurement so that incremental levels of value can then be expressed. Lord Kelvin (1824–1907) said:

> 'When you can measure what you are speaking of and express it in numbers you know on what you are discoursing. But when you cannot measure it and express it in numbers, your knowledge is of a very meagre and unsatisfactory kind.'

1.3 Value for money

Value is a comparative term expressing the worth of an item or commodity, usually in the context of other similar or comparable items. Cheapness in itself is of no virtue. It is worth while to pay a little more if the gain in value exceeds the extra costs involved. For example, it has sometimes been shown that to spend an extra sum initially on construction costs can have the effect of reducing recurring or future costs and hence the overall sum (or life cycle cost) spent on an investment. This, of course, may not always be the case since high initial costs may require high recurring costs for their upkeep or maintenance.

Value for money is an easy concept to understand but difficult to explain. It is in part subjective in its assessment in that different individuals assess different things in different ways. The appearance of buildings or engineering structures will always be largely subjective, even though a framework of rules may be devised for its evaluation. However, the opinions or judgements of others cannot be entirely disregarded. Designers have developed some rules or guidelines for the assessment or judgement of aesthetics based upon shape, form, colour, proportion, etc. The assessment of aesthetic design is difficult, since personal choice and taste are factors that need to be considered.

The engineering aspects of design, such as function or performance, are in part able to be judged against a more objective criteria. This, on the face of it, appears to be easier to assess and to make comparisons in terms of value for money. The criteria are frequently provided to designers in a client's brief and can realistically be compared with other similar structures. For example, the judgement of the spatial layout represents the adequacy of the internal space arrangement and can be related to the extent to which it facilitates the desired functions to be performed in the building. The structural components and the environmental comforts that are created can be judged in a similar manner, often using simple numerical data and analysis.

The third factor to consider is that of cost and value. The obvious approach is to put all the measurable components on one side of the equation in the form of cost, and to set these against the subjective and objective value judgements in an attempt to determine value for money. The determination of the best solution will never be an exact science, since it will always rely upon judgements set against a client's own value judgements that may be expressed as aims and objectives or outcomes. A part of value for money is quantitative in its analysis, and other aspects will always remain qualitative by definition. Value for money is the start of the process of added value. It is the principle of doing more with less – a feature that has become common in all walks of life.

As we enter the next century, a major theme remains value for money, now more appropriately described and defined as added value. Recent reports (Egan, 1998) have indicated that clients in the future will require increased value for the money that is expended on their capital projects. The principle involves reducing the relative costs of construction by designing, procuring and constructing the work in a different way than at the present time. The construction industry has, of course, been responding to this challenge with some success throughout the latter part of the twentieth century. It involves doing more for less by removing unnecessary costs. It has the aims to meet the perceived needs associated with efficiency, effectiveness and economy.

1.4 Added value

Added value is not new. It has been around for over 200 years, although until recently little seems to have been written about it. A United States of America Treasury official, Mr Tenche Cox, supposedly devised the idea. The technique is useful in measuring company performance and in monitoring manpower productivity or work outputs. It is claimed by some authors to be the real key to prosperity. It is usually referred to as added value rather than value added, since the latter description gives the impression of an afterthought, but the two terms are often thought as synonymous with each other.

Added value is a term that is used to describe the contribution a process makes to the development of its products. At a primitive level mankind goes into the forest and cuts down a tree and converts this to furniture. In so doing, value is added to the raw materials that are used. In a more complex industrial society, a manufacturing business purchases materials, components, fuel and various other services. It converts these various resources into products that can be sold for more than their combined cost. In so doing it adds value, although the amount of added value may not be significant enough to make the process worth while. Different individuals and firms will assess the significance in different ways. The construction industry uses plant, materials, people and other resources. The completed projects are usually greater than the sum of their various parts. This represents added value. Powell (1998), for example, has suggested that clients may be prepared to invest £10,000 in professional fees for £100,000 of demonstrable client benefit or added value. There is an increasing demand on the part of some clients to assess the payment of professional fees in this way. It is believed by many that where benefits, or added value, cannot be demonstrated then the associated professional services may disappear.

An alternative method of increasing added value, without changing prices or products, is to reduce the costs of the production involved. The techniques of value engineering, analysis and management have been shown to be effective in this way. The organised approach towards cost reduction goes beyond conventional methods by questioning the function of the product, component or process.

In the context of business, activities are considered to be repetitive actions that are performed to fulfil a business function. These activities can be judged to be either value added or non-value added. An activity may increase the worth of a product or service to the client, and where a client is willing to purchase such an activity it is considered to be added value. Some activities simply add time or cost to a service but do not increase its worth significantly enough to the client. In these situations they are described as non-added value. The additional time or costs here are considered to be unnecessary and should therefore be eliminated.

It should be noted that the measure of added value is not the effort that has gone into this activity. Added value is defined by the satisfaction of the customer or client and not by the producer. The manufacture of quill pens, for example, may require considerable effort on the work of the producer, but since nobody now wishes to purchase these products, no added value has been created.

An added value analysis is not merely a quantitative analysis, it is a management tool that assists in the identification of strengths and weaknesses by the interpretation of the issues raised by the added value analysis.

1.5 Efficiency, effectiveness and economy

The successful accomplishment of a task reflects *effectiveness*, while performing tasks to produce the best outcome at the lowest cost from the same resources used is *efficiency*. Effectiveness is doing the right things; efficiency is doing these things better. The best performance maximises both effectiveness and efficiency.

A building decision is effective from the point of view of the client where this achieves positive outcomes. These may be outcomes in respect of financial measures that a developer may expect, or they may be outcomes that are socially effective and may be measured using techniques such as cost–benefit analysis. In a broad sense they represent good decisions.

Efficiency results in maximising the effectiveness of a project. For example, if a building design enhances productivity while costing no more in resources than competing designs, it is described as an efficient user of these resources. Where a building project offers the same level of performance as its alternatives and costs less, then this too is an efficient solution.

In order to be effective the project must meet the objectives set by the client. Where these objectives can be achieved for less or achieve more than was expected for the same budget, then an efficient solution is achieved. Sometimes the terms *economically efficient* and *cost effectiveness* are used loosely to describe the same thing.

Economic optimisation is the process that is used to determine the most economic solution in terms of both efficiency and effectiveness. In the real world it is a goal that is rarely achievable, since possible solutions that might achieve better goals against this objective may be unknown and therefore undetected. Optimum solutions are therefore restricted to current knowledge scenarios. Optimum solutions also mitigate against the theory of continuous improvement that is a required part of added value expectations and its general philosophy.

1.6 Adding value in construction

The Quantity Surveyors Division of the Royal Institution of Chartered Surveyors (Powell, 1998) produced a report that considered the changing business world, the needs of clients and the skills required by practitioners in the future. It emphasised the importance of lean production methods and the importance of removing from the construction process anything that did not add value. This would allow a reassessment of the concepts of value and the needs of clients. In this report value was described as '*a capability provided to a customer at the right time and at an appropriate price, as defined by the customer*'. The report identified several ways in which added value might be achieved. These included:

- Facilitating earlier trading by helping clients to gain market advantage by selecting appropriate procurement routes.
- Reducing maintenance costs through the use of life cycle costing.
- Minimising disruptions to businesses during repair and maintenance works.

- Providing a point of single responsibility for clients.
- Reducing the focus on the cost of project components and increasing the focus on the benefit that the elements of the project can bring to a client's business.
- Offering a capability of calculating the added value to the client's business of different alternatives.
- Understanding how a client's profits might be increased or business costs reduced.
- Being prepared to challenge designs by suggesting new ways of undertaking work.
- Suggesting, developing and advising on different contractual and procurement arrangements.
- Understanding how construction work may be differently sourced, procured, supplied and installed.
- Gaining an understanding of how a client's competitors source work elements and carry out projects.
- Establishing sector benchmark data to facilitate inter-project comparison on design periods, lead times, installation methods and specifications for comparable projects.

In addition, a construction industry task force has issued an interim report (Egan, 1998) of its own findings and objectives. This called for:

- A reduction in capital construction costs.
- A reduction in the time available from client approval to practical construction.
- An increased number of projects completed on time and within budget.
- A reduction in the number of defects on hand-over by the contractor to the client.
- A reduction in the number of accidents.
- Increased productivity at all levels.
- Increased turnover and profitability for construction firms.

Comparisons that have already been made with other industries suggest that the targets set by this report are not overly optimistic. The increased use of information technology at all levels in industry is likely to play a major role in their attainment. The report draws on achievements now being realised in the engineering industry.

The report, *Value for Money* (Gray, 1996) highlighted the dichotomy between designs, where there is a tendency to design highly engineered, non-standardised buildings with a wealth of detail and the need for bespoke designs. This philosophy results in buildings that are complex to produce, where each building requires a new learning experience. The complexity increases costs and reduces the possibility for added value. The role of the constructor, who is possibly unknown at this stage in the design process, is not considered. The report suggested that the production-oriented approach to building design and construction that is practised in many other countries, needs to be brought more fully into consideration within the UK.

1.7 Attitude of industry towards adding value

Tables 1.1, 1.2 and 1.3 show the results of surveys of different companies (Ashworth, 1996) representing a spectrum of firms in the construction industry. The response from these firms was encouraging, in that all respondents felt that some reductions in construction costs could possibly be achieved without affecting the function and thus improve the value. However, none of those surveyed felt that a reduction anywhere near a 30% target (Construction Industry Board, 1996) was realistic, unless more radical and fundamental changes to the industry and its products were made. They also felt that such fundamental changes were likely to be unacceptable to clients. While some contractors believed that savings of up to 20% were possible, this was set against a background where they were in greater control of the building process as the lead consultant. Also in these circumstances, they expected single point responsibility and little interference or variations during construction operations on site from the client. House builders indicated that they

Table 1.1► Possible percentage reductions in cost

Respondents	Cost reduction (%)
Consultants	0–5
Contractors: General	10–20
Subcontractors	5–10
Housing	5–15
Component manufacturers	5–10
Clients	0–40

Table 1.2► Areas where savings are likely to be achieved

Area	Cost reduction (%)
Design	30
Component manufacture	25
Procurement	20
Construction	40
Management	10

Table 1.3► Some suggested areas of savings

Area	Cost reduction (%)
Design readiness	85
Single point responsibility	60
Standardisation of components	70
Standardisation of designs	45
Off-site manufacture	55
Continuity of workloads	90
Increased mechanisation	45
Teamwork	85

had already achieved considerable savings and their main problem was concerned with the lack of demand for their products and its subsequent effect upon the volume of units that were able to be constructed. Some suggested that, during the recent recession in the industry, construction costs were somewhat artificial anyway. Suppliers and subcontractors had been prepared to offer considerably lower prices for their products and services simply to maintain their business activities.

Consultants, especially designers (mainly architects), on the other hand, were rather more pessimistic of possible long-term cost reductions, although they did accept that efficiencies could always be made. They tended to suggest that as far as possible these issues were always examined carefully during the design process. There was naturally some resistance, among designers, towards further standardisation, in the use of designs and building components.

Component manufacturers were eager to find better and more economical solutions. They believed that the introduction of new technologies into the manufacturing process would not only improve product specifications but also enable their products to be manufactured more easily and economically. They were very much aligned to the progress that had been achieved in manufacturing industry in general. The only factor mitigating against this was the relatively poor state and prospects for the construction industry. This did not encourage long-term investments in expensive plant and equipment to be made. It also had a negative effect on volume manufacture and this was an important consideration in respect of reductions in cost.

While clients thought that costs could be reduced, without detrimentally affecting quality, they expected real costs to increase, especially when the recession in the industry was over. They did accept that there was always likely to be a better way of achieving the same objectives, but that consequent cost savings would be balanced against prevailing tender conditions. Many still insisted that competitive tendering provided the best opportunity to reduce costs against defined specifications in standards and quality. Any possible upturn in building activity would have the effect of artificially increasing prices. Some of the larger clients of the industry have already suggested that the '30% reduction in costs' would be insufficient to meet their own needs and expectations.

Inevitably the different organisations surveyed expected cost savings to be achieved through one of the other parties involved in the process. Consultants wanted a more precise brief from clients and better organisation from contractors. Contractors wanted more completeness in the designs with less interference from consultants, and through consultants from clients during the construction process on site. Overall, the suggestions that the scope in design cost reductions indicated that value analysis applications had still some way yet to go. The lessons suggested from buildability had not been fully realised.

All of the different parties involved wanted more ideal circumstances and situations to exist, in order to effect acceptable cost reductions. A more buoyant industry with longer term future prospects would encourage investment in the industry from all quarters. Consultants and contractors would become more profitable and this general feeling of well-being would effect greater efficiencies in practice. The stop–go nature of the industry was incompatible with achieving the best buildings. Others argued that such a requirement was incompatible with the nature of the industry and an aim that would not easily be achieved. Any attempt at smoothing out the workload pattern would be welcomed, even if this meant an industry of a reduced capacity.

Many of the specific areas where cost reductions could be achieved have been well rehearsed in recent times and are considered more fully in Chapter 2. All of the parties involved have a desire to reduce unnecessary waste and its consequent expenditure. A co-ordinated industry-wide approach would help to solve the problem in the quickest and in the most effective way. Trust between parties was a key ingredient. Some contractors argued that, in the past, some of their ideas and suggestions for reducing the costs of building and shortening the contract period had frequently been dismissed by designers under traditional contractual arrangements.

1.8 Broader issues affecting costs and values

Government

It is often suggested that the industry (a) is inefficient and clings to outdated practices and (b) fails to invest adequately due to its focus on short-term gains. The reasons for these are obvious. Those involved in the industry need to '*make hay while the sun shines*'. The booms and slumps in the industry help to encourage and prolong inefficiencies. Government could alter its own methods and timing of construction procurement. For example, it could choose not to build during times of prosperity, but only build when the industry is in recession. This would provide a dual benefit. The government would probably end up paying less for its projects and would be able to offer a more even workload to the industry. This would also go some way towards encouraging long-term investment. Instead of using the industry as a regulator for ill, it could be used for the benefit of everyone involved.

Education and training

Contractors include in their tenders substantial sums of money to rectify poor quality work on site. Training is task driven and developing the skills to carry out tasks. Education is much broader and attempts to change the attitudes and cultures of individuals. The emphasis at building craft level is now only training specific, with little attention or attempt being paid towards any aspects of education. This provides a level of skills but outside of a framework or context. The current process, however well meaning, does not engender pride in the work being carried out. This, coupled with the output-based incentive schemes, continues to a poor product, expensive remedial work and dissatisfied clients. There are exceptions; but these probably only help to prove the rule. The industry has gained a poor reputation from what was hitherto an industry in which to be proud. Better education and training could revitalise this image.

Recession

It should also be remembered that the construction industry in Great Britain, during the early–mid 1990s, had been in a severe recession – some will argue that

Figure 1.1▸ Element by element comparison of UK and US construction costs

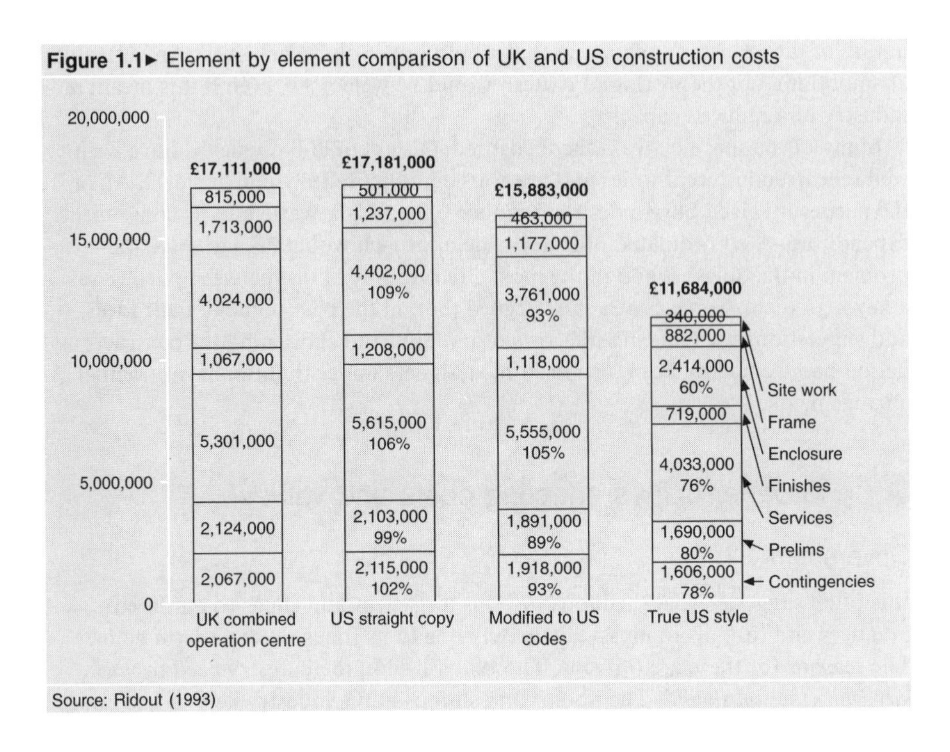

Source: Ridout (1993)

it has been the worst recession for the industry this century, i.e. worse than the 1930s. As any economist knows, prices charged for a commodity relate basically to the supply and demand of that commodity. The construction industry's products are no different in this respect. Any anticipated real cost reduction will therefore need to take into account the relatively low base of construction costs because they have been artificially repressed due to the recession. Therefore, it needs to be asked: will any cost reduction be real or only relative?

Low-wage industry

The construction industry is not a high-wage industry when compared with many others. A comparison can be made with the salaries paid at all levels in the construction industry with industries such as energy or air transport. The industry has also been described by some as a 'handicraft' industry because of its real lack of use of modern technologies. Since it is not a high-wage industry it does not encourage the best individuals to join it; at any level. For example, it has often been reported that the brightest civil engineering students do not now enter the construction professions but seek out employment in the City (of London), where salaries are higher and opportunities are better. Also, as any schoolboy knows, if you're not much good at school then you are advised to join the building industry! The good reputation and employment prospects that the construction industry enjoyed in the middle of the century are sadly not generally present today. The need for greater innovation and efficiency requires better recruitment at all levels. This inevitably means increasing salaries and wages; and hence costs, at least initially.

Inter-country comparisons

The evidence that the construction industry in other countries produces buildings more cost efficiently, and hence of improved value, than those in the United Kingdom remains inconclusive. In fact evidence might now be to the contrary. Inter-country comparisons can also be misleading (Figure 1.1). A true and fair comparison of cost is difficult to make due to a variety of different reasons, such as legislation and planning, design procedures, construction standards and quality, raw material sources, government subsidies, taxation and finance considerations. In making these comparisons suitable exchange rates, and balancing construction costs with the costs of other commodities, also need to be taken into account (Ashworth, 1999). Figure 1.1 compares the construction costs of a building in the UK with a similar building in the USA. The straight comparison reveals very little difference in actual costs. The major savings are achieved through cost reductions attributed to differences in standards, quality, regulations and user requirements as shown by the true US style comparison. Percentage comparisons are shown for the major elements.

1.9 Conclusions

There are many factors that influence the achievement of added value in the construction industry. Those employed in the industry have a large part to play and much to contribute. Government and the industry's clients also must play their part in making the industry more efficient and effective, and this should encourage more building work to take place. But this must be done at realistic prices. Added value through relative cost reduction gains are most likely to be achieved in a stable and sustained industry.

References and bibliography

Ashworth A. (1996) The shift from cost to value. *Proceedings CIB W55 Conference*, Zagreb.
Ashworth A. (1999) *Cost Studies of Buildings*, Addison Wesley Longman.
Burt M. E. (1975) *A Survey of Quality and Value in Building*, Building Research Establishment.
Construction Industry Board (1996) *Towards a 30% Productivity Improvement in Construction*, Thomas Telford.
Egan J. (1998) *Rethinking Construction*, Department of the Environment, Transport and the Regions.
Gray C. (1996) *Value for Money*, Reading Construction Forum.
Morton C. (1987) Value engineering and its importance in the construction industry. In P. S. Brandon (ed.) *Building Cost Modelling and Computers*, E. & F. N. Spon.
Kelly J. and Male S. (1993) *Value Management in Design and Construction: The Economic Management of Projects*, E. & F. N. Spon.
Powell C. (1998) *The Challenge of Change*, Royal Institution of Chartered Surveyors.
Raftery J. (1991) *Principles of Building Economics*, BSP Professional Books.
Ridout G. (1993) UK and US compared. *Building*.
Wood E. G. (1978) *Added Value – The Key to Prosperity*, Business Books.

Chapter 2

The shift from cost to value

2.1 Introduction

The practice of quantity surveying is a peculiarly British occupation and profession. It has tended to exist only in those countries where the British have had an influence in the past. It is practised in the far east in Malaysia, Singapore, Hong Kong and Australia; in India and Sri Lanka; and in several countries of Africa, such as Nigeria, Kenya, Zimbabwe and South Africa. It is also practised in Canada and in countries in the middle east that have employed British consulting engineers and designers. But just in case this sounds rather too parochial an activity, quantity surveying has been exported to countries that were otherwise thought to be foreign to the profession. It is, for example, now practised on mainland Europe and there exists *The European Technical Committee for Construction Economics*. There are a number of universities that are interested in developing quantity surveying courses in Europe and while some choose to use a different terminology the principles remain the same. It exists in the USA through *The American Institute of Construction Economics*. These are organisations largely run by and for quantity surveyors. In Indonesia the building of its airports employed a British firm of quantity surveyors. *Cost Studies of Buildings* (Ashworth, 1999) was translated into that language for local use in 1993. It must also be noted that while some countries do not use quantity surveyors as such, the costs of buildings and other structures nevertheless need to be forecasted, controlled and accounted.

According to Thompson (1968), the roots of quantity surveying go back to the seventeenth century and to the Great Fire of London. There is evidence to suggest that firms of quantity surveyors were in existence at the end of the eighteenth century. According to Seeley (1988), the earliest records of a quantity surveying firm were in Reading, Berkshire, in 1785. There is little doubt that other firms were also in existence at the same time. In 1802, a number of Scottish quantity surveying firms produced the first method of measurement of buildings. In St Luke's Gospel (14: 28) a story is recorded of the importance of counting the cost before you build, implying that some form of these practices existed even in biblical times! Prior to that time, Cheops, the builder of the world's great pyramid in Gaza, had a motto: 'I don't care how long it takes or how much it costs', but there are few in the world who can afford to have this as their aim for construction and property today. Mr Bill Gates, the owner of Microsoft, is currently building a house in the USA which at the estimate stage was calculated to cost $10 million; after partial completion, costs have escalated to $30 million. There are many examples like this that can be cited of projects that succumb to such problems, where inadequate cost control techniques are employed (Harvey and Ashworth, 1997).

2.2 Historical perspectives

The development of cost planning and cost control techniques in the construction industry can be seen through a range of different phases (Ashworth, 1983). Much of this development has occurred during the twentieth century and over the past fifty years, and can be considered in the following sequences.

Measuring

Many who performed their duties in the early days of quantity surveying had risen up from the ranks of tradesmen. Their practical skills and knowledge, together with an understanding of measurement, stood them in good stead. They were employed initially on behalf of the master craftsmen, although later they would work for the architect. Their main function was to measure the work after it had been completed and payment from the client was made on this basis. The need for change became evident as the volume and complexity of work increased and building owners became dissatisfied with the method adopted for settling the costs of the work. Also some craftsmen, for example, made extravagant claims for waste of materials. Architects, therefore then began to employ their own surveyors to contest these claims.

Documentation preparation

The principle was then established to prepare estimates for the costs of construction prior to its execution on site. There are numerous examples of projects that had to be abandoned because costs were not considered properly in advance of construction. General contractors, who were able to carry out a range of different trades, also became established during the period of the industrial revolution in the nineteenth century. They each engaged surveyors to prepare bills of quantities on which their estimates were based. The contractors realised that this was an unnecessary duplication and began to employ a single surveyor to prepare a single document on which each of the firms could then price the work. It was agreed that the contractor who was successful in winning the work would then pay the surveyor's fees. The fees were often included in the bills of quantities as a single item and paid through the contractor on the issue of the first certificate of the work completed.

Since the client was paying for this professional service, it soon became apparent that these surveyors would be better employed on their behalf under the direct authority of the architect. Surveyors later performed an independent service working with architects. They often worked directly on behalf of the client and sometimes, being the client's main professional adviser, recommended an architect to the client. Contractor's reverted to employing their own surveyors as measurers to ensure that, after completing the construction work, they received the correct payments that were due to them at the appropriate time.

The structure of the quantity surveying profession during this period was based upon a general division between 'takers-off' and 'workers-up'. In addition there was a third classification, known as the measuring surveyor. There was little doubt, however, that the important group were the 'takers-off', these being the highest paid among the salaried staff in any professional office. The aim of many young surveyors at the time was to aspire to become the chief taker-off in office. The development of these and associated analytical skills still provides the distinctive competence of quantity surveying. Accurately measuring and describing the construction work from drawings is today a much under-rated skill.

During the early years of the twentieth century, emphasis was placed on developing a method of measurement that correctly identified the costs involved in a building. In 1909, 'quantities' became a part of the RIBA (Royal Institute of British Architects) conditions of contract and in 1922 the first edition of the *British Standard Method of Measurement of Building Works* was adopted (Ashworth, 1981). Subsequent editions and revisions were prepared, culminating in a seventh edition in 1988, which was later revised in 1998. This document was part of a suite of documents developed under the Coordinated Project Information (CPI) framework by the Building Research Establishment (Willis *et al.*, 1994).

Cost planning (1960s)

It became obvious during the 1960s that the existing cost control procedures provided by the profession were too limited and not meeting the demands of the latter part of the twentieth century. Approximate estimates, which were a minor part of the quantity surveyor's work, all too often resulted in addendum bills of quantities, because the costs of the design increased beyond that which was anticipated at the client's briefing stage. There was a changing emphasis in practice from one that was largely reactive in the case of an accounting function to one that was becoming more proactive in respect of forecasting and controlling construction costs. This was the first significant shift towards adding value to the construction process.

Throughout this period those who were responsible for the commissioning of building projects had become increasingly cost conscious, in attempting to utilise scarce resources more efficiently and effectively. Changes in construction practice meant that projects were also becoming technically more complex.

Cost planning was envisaged and developed from two different viewpoints. One sought to limit the total costs of building and was referred to as elemental cost planning or *designing to cost*. The other, developed by the Royal Institution of Chartered Surveyors Cost Research Panel, was referred to as comparative cost planning or *costing a design*. The two systems were eventually developed into a single system utilising the best attributes from each (see Chapter 4).

Graduates and technicians (1970s)

During the 1970s, a radical shift in education and training took place among potential surveyors (Ashworth, 1994). Historically, new recruits to the profession had, after leaving school, found appropriate employment in surveying practices. Professional status had been achieved through taking the Institution's examinations, often after studying on a part-time basis at a local college or through a distance learning course.

The professional institutions had, during the mid-1960s, increased their academic entry qualifications to two General Certificate of Education Advanced level subjects. This was comparable to entrance to university. Since less than 10% of the population then went to study at university, surveyors easily fit within that profile. Government policy of the day also reflected a desire to increase yet further the overall level of qualifications among all young people, and there was therefore an expansion at all levels of education, including additional compulsory education at school. There was also an expansion of technician education, aimed at developing a two-tier system of employment in practice among the many different professions.

The implications for quantity surveying were that the graduate quantity surveyors would become more concerned with advising clients on the broad issues and applying basic principles to solve new problems. The type of education would also provide a more enquiring and educated mind by which to question and improve the quality and type of service provided. The technician surveyors would largely undertake the main core of activity, such as document preparation, interim payments and the preparation and settlement of final accounts. The graduates would quickly develop to oversee these activities and be responsible for the strategic matters associated with their professional practice.

While the different roles continue to exist within most practices, the demarcation between the work of graduates and technicians varies greatly between one practice and another. At best it is blurred. There was also the aspirations of the technicians to be met, with many of these wanting to move upwards towards the professional roles. Later this would be achieved through the introduction of undergraduate degrees by part-time study.

Another spin off for the profession of such a change in education policy was the number of quantity surveyors who became employed as full-time lecturers in education. Within the United Kingdom alone this figure at one time exceeded 200, with perhaps, in addition, as many as this elsewhere throughout the world. This allowed research and scholarly activity to be developed, not solely by these surveyors, but also through practice and by those in practice studying for higher degrees. This further enabled the profession to become better prepared for the future and allowed new development opportunities to be explored. However, it has to be recognised that the majority of new developments in practice have evolved or have been developed directly by those employed in practice. Educational research offered a new perspective in terms of explanation, understanding and was not so restricted by the commercial environment. There have often been tensions between practitioners and researchers, although today there is frequently a closer working relationship to secure common objectives. The comment made by a former president of the RICS (Townsend, 1984) about the lack of links between academic research and the needs of practice could not now be fairly applied.

Management and modelling (1980s)

A further shift occurred in the way in which projects were organised. New procurement methods were constantly under review, design and construct became more popular, subcontracting took over from general contracting, construction processes and procedures developed a more international flavour, the industry became too adversarial, and information technology began to be harnessed for advantage.

Project management began to be seen as a skill that clients would require in the future and several practices began to offer this service in addition to their more defined activities. The National Exhibition Centre in Birmingham, which had been project managed by a quantity surveyor, provided a good advertisement for their services.

Quantity surveyors have also been involved in several different management roles, on behalf of both clients and constructors. Many large contracting firms, for example, have, as their chief executives, individuals who have previously trained as quantity surveyors.

While the bill of quantities can be described as a pragmatic model of construction costs, it was developed as means of expediency, rather than on the basis of any analysis of the way that costs are derived. Its existence has for many years been cast in doubt and its long-term demise is supposedly getting closer. It has been threatened because of its own limitations, changes in procurement and contracting and more recently through the advancements in information technology and the rapid developments in computer-aided design.

Some attempts have been made to model costs in different ways through statistical techniques, simulation and expert systems (Skitmore, 1999). While costs will always remain an important ingredient in building design, there are likely to be different and more efficient ways for deriving its calculation. The development of various types of cost modelling was, and still is, seen by the profession as too radical, requiring practice to change from the security of a reasonably reliable but yet imperfect system. It would like to improve the reliability of its processes, but does not yet see in new modelling techniques a valid way forward.

The emphasis on building costs also switched from that which concentrated on initial costs alone towards a more holistic view of costs over the entire life cycle of the project. Building costs were also contrasted with the other costs associated with development such as land and furniture and equipment. Eventually all of these aspects would be considered within the emerging practice of facilities management. These considerations provided yet another shift towards adding value to the industry and its clients. This recognised that to examine the initial costs of construction within the context of whole costs, might yield an overall better financial solution for the owners and users of property.

Diversification (1990s)

During the 1990s, several key reports were produced that have assisted in the new direction of the professions. These helped to formulate a future direction and strategy, emphasising the changing nature of society in general and the construction industry and the profession within it. It also recognised the future importance that should be attached to research and innovation and an appraisal of the skills and knowledge base of the professions. The role of the quantity surveyor has changed considerably during the past fifty years. This will probably be over-shadowed even by the anticipated changes that might be expected in the future. We often overestimate what will happen in the next two years, but underestimate what may happen over the next ten years. The following are some of the issues which will face the profession at the beginning of the new millennium.

Information technology

The use of computers for preparing bills of quantities in the late 1960s has moved to such an extent that the capability of converting designer's drawn information into a contractor's tender at the press of a button is now achievable in practice. It relies upon the designer producing appropriate drawings in the first place, but even in the absence of these the computer is able to make assumptions that can be easily changed to suit the needs of a design. Where the computer link between drawings and quantities has not been made because of an architect's preference for a particular CAD system, the link between quantities and a contractor's tender is now being used in practice. EDICON (the UK construction industry forum for electronic data interchange) has developed a system which meets this capability. Integrated packages that link pre-contract documentation and post-contract work are now available. Greater use in the future will be made of those computer systems that capture the expertise of the practitioner and refine it for future applications. Information technology allows decisions to be made using better quality information, providing yet further evidence of adding value for the industry's products, processes and its clients.

Professional boundaries

There has been a blurring of professional boundaries over the past few years. The quantity surveyor's role, as in many other professions, has been to diversify into work that previously would have been undertaken by another professional discipline. This is true of all professions both within and outside the construction industry. The management consultant has arrived: a person who is able to solve a client's problem or employ the services of someone who will be able to offer the appropriate advice. The concern of many large surveying practices is that they are no longer the preferred lead consultant on a construction project. Clients who have formed strong links with accountancy practices through auditing now turn to such firms to advise them on their construction proposals.

QS 2000 (Davis, Langdon & Everest, 1991) stated that '*Significant changes have been occurring in the structure of the profession as a result of wider changes in the industry.*' Significant, but less measurable, are changing attitudes to practice and professionalism among, in particular, younger quantity surveyors responding to the more aggressive and commercially minded working environment in the mid-to-late 1980s. It is clear that the role of the quantity surveyor is expanding, and this is reflected in the growth of the bigger interdisciplinary and international QS practices as well as the growth of niche practices specialising in, for example, taxation advice or dispute resolution.

Business orientation

According to QS 2000, '*practice is increasingly characterised by a business oriented approach emphasising, for example, rapid turnaround of information and improved quality of communication and presentation*'. There remains a dichotomy between the definition of a business and a profession, and this is mirrored elsewhere among all the major professions. The driving force in the past was to put the quality of service and professional advice above profits. Business practice tends to turn this approach on its head.

International

It has already been suggested that quantity surveying is a peculiarly British profession that has influenced those countries with former links through the Commonwealth. It has also been stated that quantity surveying exists in many countries abroad, over 100, as stated in *The Future Role of the Chartered Quantity Surveyor* (RICS, 1983) twenty years ago. While the integration of Eastern Europe into Western ideas has opened up new opportunities, quantity surveyors have been working in Western Europe for many years. The single European market which came into operation on 31 December 1992 offered an additional opportunity for quantity surveying skills and practices.

In addition to Eastern Europe, China, South America and South Africa are seen to offer huge opportunities for quantity surveyors, well into the next century.

Quality

The 1990s were labelled the decade of quality, with many companies seeking to demonstrate that the services that they provide are within a quality framework, such as ISO 9001. This has become an issue for quantity surveying practices that have not yet established systems to ensure that the quality of the services provided fits within a defined specification. Procurers of professional services are looking towards recognised kitemarks or similar systems that will provide for them an assurance on quality performance of a company.

Knowledge explosion

Knowledge is continuing to increase at a pace. This is evidenced all around us in an age of knowledge explosion. It is estimated that as much as 50% of all world-wide knowledge has been acquired during the latter half of the twentieth century. Information, it is claimed, is doubling every 73 days! Coupled with this, there is a better understanding, through research and development, of processes and procedures associated with cost studies of buildings, and the percentage gain is from a low base level and therefore much higher in this subject area. It is not just a question of achieving a know-how, but also in developing some understanding of a know-why. This can be directly attributed to increased levels of research activity. At the same time new skills are being acquired for new applications in what has become an age of rapid change.

2.3 Strategic directions

QS 2000: The future role of the chartered quantity surveyor

The theme of this report (Davis, Langdon & Everest, 1991) was change: changing markets, changing industry, changing profession, changing society. The report identified the challenges that faced quantity surveying, together with the opportunities for future roles and levels of activity. The importance was stressed in relation to managing time, cost and quality more effectively and in adding value to the client's business and construction project. This further emphasised and reflected upon the shift from cost to value, which is now well recognised. QS 2000, on the issue of bills of quantities, did not expect them to disappear but to become increasingly more automated. Perhaps bills are now embedded in the British way of life?

The report identified widening markets and diversification for the quantity surveyor, suggesting three key areas of future activity: value management, procurement management and facilities management. New skills, knowledge and understanding would be introduced progressively stemming from research and development. These would be an aid for adding value for clients.

The core skills and knowledge base of the quantity surveyor

This report (RICS, 1992) identified a framework for the classification of skills and knowledge that professionals possess. It suggested that knowledge develops from many different sources and continues to evolve throughout the life of an individual, but more quickly during the time of intensive study.

The report identified the key areas in which the quantity surveyor's skills lie. It suggested a ranking for these skills with implications for education and training. It differentiates between techniques and skills by suggesting that the latter are taken to be the ability to apply the techniques effectively and efficiently to a range of problems encountered in quantity surveying. Underlying all quantity surveying, is the general ability to think clearly and methodically and to be able to communicate to both experts and lay people alike. The skills are identified with the individual's skills as well as the collective ability of the profession as a whole. These include: management, quantification, documentation, analysis, appraisal, synthesis and communication.

The report also considered the ethics or constraints which professional people impose upon themselves and the relationship between commercial and professional roles. The ethical constraints are determined not only from the individual behaviour of surveyors, but also through the imposition of the by-laws and codes of conduct of the professional bodies. Other constraints may arise due to a lack of understanding from clients but they also relate to the individual capabilities of surveyors or practices.

The challenge of change

This report (Powell, 1998) sought to find solutions to questions such as:

- What do customers want?
- What will the markets be like?
- How can the value of services be improved?

It focused on developments such as the business world and its effect upon the construction industry. It emphasised the global perspective and considerations of many of its customers, with expectations of lower costs, higher quality with speed and safety – a simple definition of added value. It also considered the implications from information technology and a recognition of life-long learning as being fundamental to career prospects. The report identified several areas where the quantity surveyor could add value, but emphasised the importance of fully understanding a client's business objectives. The report fully recognised the importance of maintaining or increasing function, while at the same time reducing cost (i.e. adding value).

Research and innovation

The importance of research and other scholarly activities is now recognised as an integral part of the developing evolution of the surveying profession. In *The Future Role of the Chartered Quantity Surveyor* (RICS, 1983) it was stated that '*unless an academic discipline is related to a substantial activity in the field of research and development, it simply becomes a process of instruction by rote*'. The importance of research in education and training cannot therefore be over-emphasised. An HMI (Her Majesty's Inspectors) report (Department for Education, 1992) further stated, '*that research and scholarly activities underpin curriculum developments*'. The Centre Scientific et Technique du Batiment (Kennaway, 1984) commented: '*the stronger the Research and Development effort of a sector, the better its image; even in a fragmented sector. Look at the image of doctors!*'

The commitment to research that is aimed at determining the future shape and direction of the profession is important. Research topics, identified by Brandon (1992), include a variety of subjects that are of importance and of relevance to the practice of quantity surveying: development appraisal, value management, estimating and bidding methods, risk analysis, life cycle costing, expert systems, computer-aided design, integrated databases and procurement systems.

A report published by the Royal Institution of Chartered Surveyors titled *The Research and Development Strengths of the Chartered Surveying Profession: the Academic Base* (RICS, 1991) undertook a survey to help to determine the extent of these activities in the universities. The information included an overall profile of research activities, general areas of capability and specific research expertise, research links with the profession and details of external research contracts.

The importance of research is in adding value to the process by either doing things better, doing new things or through gaining a better understanding of what is trying to be achieved.

Table 2.1▸ What the private sector client wants: industry performance compared to the car industry

Wants	Modern motor car	Modern buildings		
		Domestic	Commercial	Industrial
Value for money	••••	•••••	•••	••••
Pleasing to look at	••••	••••	•••	•••
Free from faults	•••••	•••	•	••
Timely delivery	••••	••••	••••	••••
Fit-for-purpose	•••••	••••	••	•••
Guarantee	•••••	••••	•	•
Reasonable running costs	••••	••••	••	•••
Durability	••••	•••	••	••
Customer delight	•••••	•••	••	••

Source: Presentation by Dr Bernard Rimmer, Slough Estates plc, to a conference organised by *Contract Journal* and CASEC, the Barbican, London, 15 December 1993

2.4 Constructing the team

The construction industry has, over the past fifty years, produced several different reports, many of which have been government sponsored. Their overall themes have been aimed at improving the way the industry is organised and the way construction work is procured. The underlying theme in all of them has been to attempt to provide better value for money for the clients or customers of the industry. The Report Latham titled Constructing the Team (Latham, 1994) is just one of many.

The report is a review of the industry. It suggests, for example, that when buildings are compared with the motor car, the result is unfavourable (see Table 2.1). While the automobile industry has made huge strides with the quality and reliability of its products, the same cannot be stated about the construction industry. The report further suggests that the way in which the industry is organised contributes towards this poor view. This is also often a view that is expressed by members of the public. The report makes reference and recommendations on:

- Use of co-ordinated project information.
- Importance of construction in the wider economy.
- Need to reduce adversarial problems.
- Public image should be improved.
- New research initiative should be formulated.
- Effects of ISO 9000.

An item of major importance to the quantity surveying profession was the suggestion that a 30% real cost reduction by the year 2000 should be targeted by government ministers and the construction industry. These cost reductions should not generally reduce quality but should at least maintain but preferably improve the quality of buildings, noting the comments expressed by clients and the comparison with the motor car in Table 2.1. However, some major clients of the construction industry have suggested that such a cost-value reduction would be insufficient to meet their overall aims and objectives. The implication is one of adding value.

The report suggested the establishment of a Construction Development Agency with one of its key tasks to reduce the costs of construction in the United Kingdom. The Construction Industry Board (1996) published a report that sought to explain how this might be achieved. Other publications have also sought to identify how such improvements could be made (RICS, 1995; Gray, 1996; Ashworth, 1996). Ashworth has also sought to identify the responses from industry on how such cost reductions might be achieved by identifying areas of possible savings in costs and prioritising these in their order of importance. This process was one of adding value.

It is also very important that such cost reductions do not refocus the construction industry backwards fifty years towards the emphasis on initial costs. The importance of ensuring that life-cycle costs are given their rightful importance in the overall building process must be maintained.

2.5 Other factors to consider

Before considering areas where added value is possible, the following points should be borne in mind. It must be accepted that substantial cost efficiencies have already been introduced and maintained over the last fifty years in many different ways through, for example:

- Cost planning
- Mass production and bulk purchase
- Prefabrication
- Buildability
- Subcontracting
- Off-site manufacture

This, of course, in no way supports a view that further cost-value reductions are not possible and that added value is not attainable. In fact it strengthens the argument of seeking out such efficiencies. The following are some examples of the possible areas associated with design and construction that could be examined for possible cost-value reductions.

Organisation

- Planning and Building Regulations requirements.
- Procurement arrangements.
- Government stop–go policies.
- Materials imports.
- Use of information technology.

Design

- Over-specification in some areas.
- Incomplete design at tender stage.

- Separation of design from construction.
- Off-site prefabrication.
- Standardisation of components.

Construction

- Construction as a manufacturing process.
- Getting it right first time (avoiding defects).
- Wider use of mechanisation.
- Better training of operatives.
- Reduction of labour input (see below).

Manufacturing industry succeeds where it is able to produce a large number of standard units or components. Compared with thirty years ago many of the construction industry's products are now manufactured off-site under controlled mass-produced conditions. The single difficulty that exists with these sorts of products is that when minor repairs are necessary, they often require replacements of components far in excess of the immediate problem with consequential costs to follow. This is a factor that must be considered in the life cycle costing of projects. Gray (1996) comments that research has shown that the main cause is the national tendency to design highly engineered, non-standard buildings with a wealth of detail. This results in buildings that are complex to construct, with each building requiring a new learning experience.

Construction as a manufacturing process

Manufacturing industry has realised that the expensive component in their production is labour. In fact it's the most expensive part of any production process. The examination of any photographs of manufacturing workshops at the turn of the century confirms the startling contrast between then and now in the number of people employed. Construction work on site, on the other hand, still remains very labour intensive. It has not realised the benefits of applied information technology in, for example, the numerical control of plant and equipment for routine site operations. While there is undoubtedly more controlled off-site manufacture, requiring assembly only on site, the process of erecting buildings has otherwise changed very little. The development of mechanisation on site has evolved to only a limited extent over the past fifty years. The new technologies that are being routinely used in manufacturing, such as computer-controlled machines, have had only limited impact on the processes used on the typical construction site. The machines used in manufacturing not only produce goods to a predefined standard but also have assisted in reducing some of the major costs associated with manufacturing production.

Materials and components

A number of manufactured components are still imported from overseas countries. Sensibly, for a country as a whole, emphasis should be given to producing materials, goods and components within the country's own economy to the appropriate standards and costs. Such costs are then much more controllable. The importing of such items must inevitably increase the respective costs to the end user. In a country-wide comparison, this might indicate that the use of overseas products is increasing the costs of United Kingdom construction unfairly.

Procurement

A lot of attention has been given towards the different procurement methods used in the construction industry. Research has largely been inconclusive regarding the optimum choice of method to provide the best project overall, either in terms of cost, quality or design.

Construction is also one of the few industries that continue to separate design from the production process. However, this is also one of its distinctive features, recognising that the construction of buildings and other structures is often bespoke to a client's needs and represents an inherent difference from other industries. The production of 'standard' buildings that require limited architectural intervention, provides no real argument for the combination of design and construction generally.

Unlike manufacturing, the production of construction projects on site requires variations. The reasons for these is that the typical construction project is not fully designed prior to starting work on site, which inevitably increases inefficiencies. A study of hospital buildings, conducted several years ago, revealed that if the project was fully designed it could be constructed more quickly on site. A shorter more effective time on site usually means a saving in costs. By adopting such an approach, the overall time reduction between inception and completion could also be expected, resulting in a much earlier hand-over to the client.

2.6 Conclusions

The quantity surveying profession has evolved a long way since it was first conceived over 250 years ago. It was then a post-measurement and accounting discipline only. With the increased emphasis placed upon the costs of building and the importance of adding value, the role of the quantity surveyor has increased in importance.

It began to develop methods that would assist with the forecasting and controlling of costs. These costs were generally interpreted to be initial costs of construction alone, with no reference or activity associated with costs-in-use. The interest in whole life costing or life cycle costing during the 1980s expanded the work and the role of the quantity surveyor. The continued importance of the economic choices associated with limited resources has assured quantity surveyors of a prominent future in the construction industry. However, they need to expand their skills and expertise in order to better inform and advise clients. The emphasis in their work in the future will be in improving or adding value to the construction process by seeking out ways of doing more for less.

References and bibliography

Ashworth A. (1981) The SMM story. *QS Weekly*, January.

Ashworth A. (1983) Fifth generation surveyors. *Chartered Quantity Surveyor*, September.

Ashworth A. (1994) *Education and Training of Quantity Surveyors*, Chartered Institute of Building Construction Papers, No. 37.

Ashworth A. (1996) The shift from cost to value. *Proceedings CIB W55 Conference*, Zagreb.

Ashworth A. (1999) *Cost Studies of Buildings*, Addison Wesley Longman.

Brandon P. S. (ed.) (1992) *Quantity Surveying Techniques: New Directions*, Blackwell Scientific Publications.

Construction Industry Board (1996) *Towards a 30% Productivity Improvement in Construction*, Thomas Telford.

Davis, Langdon & Everest (1991) *QS 2000. The Future Role of the Chartered Quantity Surveyor*, Royal Institution of Chartered Surveyors.

Department for Education (1992) *Built Environment Education in the Polytechnics and Colleges*, Department for Education.

Gray C. (1996) *Value for Money: Helping the UK Afford the Buildings it Likes*, Reading Construction Forum.

Harvey R. C. and Ashworth A. (1997) *The Construction Industry of Great Britain*, Butterworth-Heinemann.

Kennaway A. (1984) Errors and failures in building: why they happen and what can be done to reduce them. *The International Construction Law Review*.

Latham M. (1994) *Constructing the Team*, HMSO.

Powell C. (1998) *The Challenge of Change*, Royal Institution of Chartered Surveyors.

RICS (1983) *The Future Role of the Chartered Quantity Surveyor*, Royal Institution of Chartered Surveyors.

RICS (1991) *The Research and Development Strengths of the Chartered Surveying Profession: The Academic Base*, Royal Institution of Chartered Surveyors.

RICS (1992) *The Core Skills and Knowledge Base of the Quantity Surveyor*, Royal Institution of Chartered Surveyors.

RICS (1995) *Improving Value for Money in Construction*, Royal Institution of Chartered Surveyors.

Seeley I. H. and Winfield R. (1999) *Building Quantities Explained*, Macmillan Education.

Skitmore M. (1999) *Cost Modelling*, E. & F. N. Spon.

Thompson F. M. L. (1968) *Chartered Surveyors: The Growth of a Profession*, Routledge & Kegan Paul.

Townsend G. (1984) Efficiency and profit. *Chartered Quantity Surveyor*, December.

Willis C. J., Ashworth A. and Willis J. A. (1994) *Practice and Procedure for the Quantity Surveyor*, Blackwell Scientific Publications.

Chapter 3

Design and construction

3.1 Introduction

The historical development of the structural form is one of slow growth, of experiment, success and failure. Occasionally a unique achievement proclaims the limit of a particular structural development. More usually the limitations of available knowledge, skills and suitable materials restrict this natural evolution. Within the last century, steel, reinforced concrete and aluminium have brought about a revolution in structural design. This has been achieved not only through the properties of the materials themselves, but also through increasing knowledge and experimentation.

The design of buildings is influenced by a combination of different factors such as:

- The physical characteristics of the site.
- The period in history when the projected is constructed.
- The culture prevailing at the time.
- The material resources available.
- The building techniques and processes that can be used.

The three factors that govern the form of any building are its purpose, the materials to be used and the skills of the constructors. Buildings (architecture) were perhaps best defined by Mies van der Rohe when he said: '*In its simplest form, buildings (architecture) are rooted in entirely functional considerations, but they can reach up through all degrees of value to the highest sphere of all, into the realms of pure art.*'

3.2 Social contexts

English architecture may be broadly described into three main periods. Within these periods further subdivision is also possible. There were also periods of transition.

Medieval

This period extended from the Dark Ages to the Reformation, i.e. from 600 to 1500. It was by far the longest of the three. Men were influenced as never since by a tremendous religious preoccupation. During this time, therefore, the buildings on which men lavished their greatest skills and care were those of a religious nature. Ninety-nine per cent of the buildings that survived from that time are churches, cathedrals and monasteries. This indicates that these were the types of building that would endure over time, and nothing was spared to make them beautiful. The remainder are castles that were built to subdue the country and to house the feudal lords.

Renaissance

The second period covers a time when intellect was considered to be all-important. Religion had suffered a remarkable decline and social organisation had advanced. The period lasted for no more than about 200 years. In these circumstances, material comforts and improvements in domestic building were natural – a process that was greatly helped by the rapidly increasing wealth of the country.

Industrial

This is the period during which men were building for gain. The industrial era cannot be said to have started at any particular date, but the final ten years of the eighteenth century and the first seventy-five years of the nineteenth century constitute a period of industrial growth that proceeded at a frightening pace. This was largely made possible by the fact that coal and iron were found together in many different parts of the country.

Society moved from one that had a reliance on agriculture to one that was based on manufactured goods from factories in amounts that seemed to double or treble each year. The increase in goods traffic was enormous; elaborate canal systems were first built and then railways were laid down. Business increased with manufacture, so huge offices and banks were required. Shops and warehouses were constructed as the country moved more towards a consumer society.

During the nineteenth century, the principal types of buildings were industrial, commercial and civic. During the twentieth century the accent switched towards domestic buildings and the improvement of public buildings.

Until the end of the eighteenth century an ordered existence had prevailed. In the nineteenth century, life became chaotic as developments were increasing at such a rapid pace. In attempting to rectify these affairs, planners were evolving to control the location and type of development.

3.3 Design

Following the ending of the Second World War, most European countries found themselves poorer. Some, like the United Kingdom, were also physically devastated. During the post-war period of reconstruction, architects and engineers strove to harness the industrial production techniques developed during the war for civil construction programmes. But the balance between expediency, easily borrowed from the military culture, and the requirements to provide a humane environment were not easily accomplished. Many of the 'factory-system' building methods developed during the 1950s and 1960s in Scandinavia, France, Germany and Britain produced buildings that were very dismal in appearance. This was especially true when such methods were applied to large-scale housing developments. The seemingly endless prefabricated concrete wall panels that were stacked together in walls or towers came to symbolise a paucity of imagination, stringency of resources and the careless attitude of politicians (Fletcher, 1996).

During the twentieth century, and in order to break away from the applied period styles which had become accepted as the meaning of architecture, designers reverted to first principles and a dogmatic adherence to the functionalist ideal. They believed that if a thing was truly fitted to its purpose then it must necessarily be beautiful. This was in contrast to the revivalists who believed that beautiful buildings could only be so if they were built in the manner of a past age. However, functionalism alone could not produce great architecture since it overlooked important factors such as form, texture and proportion.

The needs of today are very different from those of the past, and our scale of values has undergone a profound change during the twentieth century. For example, we need different sorts of buildings that did not exist previously, such as health centres, airports, flattened factories for modern methods of production, television studios, etc. The impact of the motor car on our environment has in itself been considerable. The problems of how to deal with the motor car without destroying the character of our old towns and cities has still to be resolved. Where whole new towns have been developed comprehensively, there has been no reason for not integrating this aspect.

Modern methods and industrial techniques have increased the variety of materials and products that are now easily available. To the natural materials have been added steel and concrete (the most important), plywood, plastics, aluminium, bitumen and many others. Old materials, such as glass, have found new forms in plate glass. The increased standardisation of some building components have allowed them to be delivered directly to the construction site for immediate installation.

The skills used today are a mixture of specialist and multi-skills where the emphasis on craftsmanship has partially moved from man to the machine. Compare, for example, the skills required today in a modern complex building, with its high reliance on engineering services, with those of a medieval church of the same capacity, in which the materials used were limited virtually to stone, timber, wrought iron, glass and lead. Today many of the building trades are involved in the assembly of components on site, requiring different kinds of skills.

3.4 Technology

Changes in construction technology continue to take place due to research, innovation and developments from practitioners.

Off-site manufacture

There has been an increase in the practice of manufacturing a wide variety of building components off-site under factory conditions and transporting them to the site to be incorporated in the construction. The advantage of the controlled environment is that better quality control can be achieved when compared to work on site. Site labour is saved, but this saving is offset by the extra factory manpower required, nonetheless resulting in an overall reduction in cost. Under factory controlled conditions such components can be more accurately cast or fabricated. Handling, transportation and component assembly in the project are now important considerations at the design stage. The use of prefabricated components has helped to contribute to the overall levels of improved productivity. The characteristics of conformity and simplification, together with the economies of scale achieved through volume production, provide a combination of economy and speed of construction. The introduction of off-site manufacture has had implications for the skills and attributes required from site operatives in terms of the site assembly of the components.

Intelligent buildings

Business in the 1990s often requires buildings to be intelligent or, in United States terminology, 'smart'. An intelligent building is one that satisfies the requirements of its occupants. This might include flexibility which caters for different work groups with varying needs at locations that may be subject to frequent change. The latter factor is often referred to as the churn rate, i.e. the number of people moving location within a building each year, which can approach or even exceed 100% (Harvey and Ashworth, 1997). The operation of business is dependent on massive information technology investment, which facilitates information exchange within buildings and between buildings. This is more than enhanced by telecommunications; these services include word processing, electronic mail, video conferencing and networks which access a wide range of business databases. Electronics contribute to the handling of the environment, security, the regulation of access including lift movements, energy and lighting management and operation. Futuristic concepts include structures that can respond in a more sophisticated way to the changes in the external environment using insulated glass with variable opacity (i.e. smart glass), materials with variable thermal capacity, self-cleaning and maintenance services.

Building change needs management. The more intelligent the building, the more difficult it is to manage. To reorganise heating, cooling, lighting, security and other facilities requires more sophisticated systems and highly trained personnel to carry out their operation. To provide advice and guidance to these personnel an expert system can be used. Indeed, there is a view that an intelligent building is one with an expert system incorporated to facilitate building management. The Building Research Establishment has developed an expert system (BREXBAS) which monitors sensor data from remote systems, applies its knowledge and reasoning capabilities to interpret the information, and generates advice for the user. Greater control is gained over increasingly expensive building space, enabling it to be used more effectively and efficiently.

Robotics

Construction work requires the execution of dangerous and demanding tasks in an environment which can be dark, dirty and, in the longer term, injurious to health. In order to relieve construction workers from the rigours and hazards of harsh operational environments there is potential for the development and application of robotic devices. Such devices have been used successfully in the nuclear and steel industries, but construction has not yet fully benefited from these technological advances. With these applications, it is expected that there will be the additional benefits of increased productivity together with improvement of quality. These advantages have been seen more clearly in Japan where the development of construction robots is most advanced. Brown (1989) cited 89 examples of construction robots of which 74 were Japanese. Each of the top six construction companies in Japan has produced working prototype robots, although whether they are yet fully cost effective for normal site operations, even in a land where site labour is highly paid, is questionable. Nonetheless, they offer considerable time savings when their output is compared with that of a skilled operative. These include robots for:

- The erection of steel structures.
- The finishing off of concrete floors.
- The site welding of steel components.
- Painting.

A frequent objection to the use of robots is that the construction site is not orderly, as is a manufacturing operation, but is constantly changing and the multiplicity of craft operations that are carried out are not suited to replacement by robotic devices. Notwithstanding these objections, there have already been a number of successful robotic devices used in construction and there is little doubt that these developments will continue even though, at this stage, it is not possible to define what their developed forms may be.

Performance monitoring

Until fairly recently the construction industry, unlike the engineering or chemical industries, paid relatively little attention to the performance monitoring of its artefacts. Although it was recognised that the deterioration of a structure commenced during construction and continued with time, even maintenance was undertaken only when structures showed signs of distress. However, the advantages of maintenance and performance monitoring have now been recognised (RICS, 1990) and greater attention is being paid to the monitoring of building structures using surveying and photogrammetric techniques, dynamic testing and automatic monitoring (Moore, 1992). To counter rapid deterioration of buildings it is essential that the contributions of the designer, building contractor and building surveyor are properly undertaken. Yet there is no basis for allocating the balance of expenditure on design, checking, supervision and inspection. There are no cost-benefit systems for error control. Standards are set for the appropriate levels of materials quality and workmanship without a clear understanding of how this impacts on the safety of the structure. Performance monitoring:

- Confirms that the behaviour of the structure is in accordance with the predicted behaviour.
- Gives knowledge of any discrepancy in performance.
- Gives an opportunity for corrective action.
- Improves the body of knowledge that is required to undertake design with confidence, thereby improving future structures.

Offshore structures and construction activities have, in particular, been subjected to close performance monitoring as the detection and repair of fatigue damage is essential. There are considerable uncertainties in the prediction of fatigue lives and scope for the development of a reliable theory to indicate the relative risk levels of fluctuating loading. The development of performance monitoring techniques has created an alliance between the construction industry and instrumentation engineers.

Sustainable development

The impact of the construction industry on the environment is substantial. During the extraction and manufacture of construction materials, large quantities of energy are used in their transportation, the process of construction and the use of the buildings. Major contributions are made to the overall production of carbon dioxide which exacerbates the 'greenhouse' effect. The environmental impact is global but, during the construction process, communities and individuals are affected.

Society is becoming increasingly concerned with the effect of human activity in the environment. In recent years there has been greater pressure on clients to state all the likely direct and indirect effects of their projects on the life and amenities of surrounding areas. It has been the practice to supply environmental impact assessments with planning applications for major projects. The assessments require a statement of the impact of the project on the surrounding area. The statement should also include details of work which will limit the impact – for example, soundproofing, as in the case of a noisy transportation system. To involve the public more closely the assessment statement should be jargon free. A further requirement is that promoters should consider the detailed impact of their project early in the planning process and undertake wide consultations involving the public and environmental groups. Despite the requirements specified by the directive, there is concern that there is a lack of definition regarding the scope of the assessments and the level of detail required. Assessments frequently tend not to look beyond the confines of the project concerned.

The concept of a green building is an elusive one. The definition is broad and being green in a professional sense may merely come down to a change in attitude. Most buildings in the UK are designed to cope with the deficiencies of a light, loose structure, designed to meet the Building Regulations' thermal transmittance standards and no more. As roughly 56% of the energy consumed, both nationally and internationally, is used in buildings, this should provide designers with opportunities and responsibilities to reduce global energy demand. There is a need to make substantial savings in the way that energy is used in buildings, but there is also a need to pay attention to the energy used in manufacture and fixing in place of a building's components and materials. For a new building this can be as high as five times the amount of energy that the occupants will use in the first year.

3.5 Design and technology

Changes in the evolution of building design and technology can perhaps be best illustrated by comparing two domestic buildings, one designed at the turn of the twentieth century and the other at the turn of the twenty-first century (see Figure 3.1). This evolution has involved the introduction of new ideas in construction technology and materials. This is in response to more carefully evaluating what is provided, at what cost and how this might be achieved more economically, thus adding value to the project. The introduction of new materials can sometimes be traced to difficulties in the supply of traditional materials, their expensive costs or their poor reliability in use.

Figure 3.1► Traditional and modern design and construction

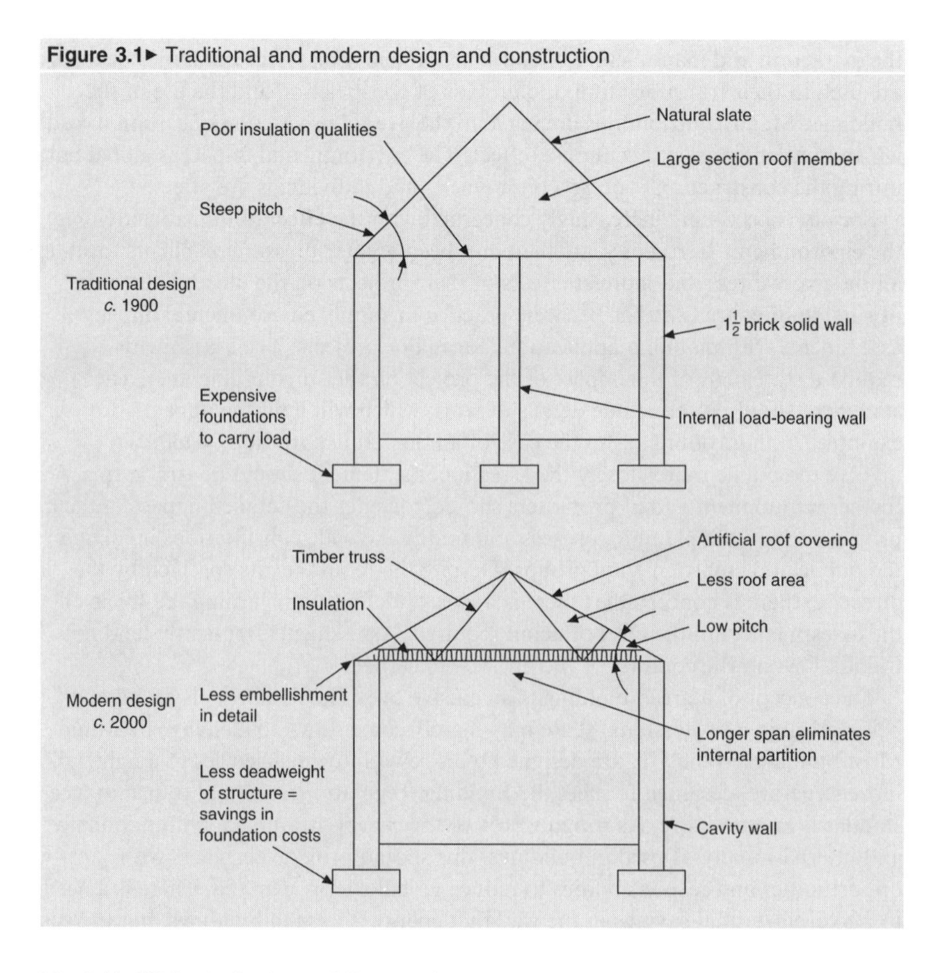

The roof construction of buildings is always worthy of study, since on all projects, other than multi-storey construction, roofs can be described as a cost-sensitive building element. The cost sensitivity of an element is dependent upon its total cost compared with the total cost of the building. In adopting alternative technology and materials, the point needs to be understood that the architecture, in many respects, remains unchanged. For example, in Figure 3.1 only the pitch and the roof covering appearance are different between the two designs. In terms of aesthetics, the lower pitch may be more desirable, depending upon one's taste. It is also becoming possible to create artificial materials that are now comparable in performance and appearance to the much more expensive traditional and natural materials. Building performance and function have improved. More attention is also being given towards design life – a concept that was not really considered in 1900. Introducing materials that have life expectancies well beyond that of the building is wasteful, even where the materials may be partly reclaimed on demolition.

The comparisons between the designs in Figures 3.1 yield the following added value in favour of the later building:

- Lower pitch requiring a smaller area of roof covering, shorter lengths of rafter, etc.
- Lighter weight artificial materials offering a similar appearance and other performance characteristics at a lower initial cost.
- Use of prefabricated roof trusses, offering smaller sectional sizes of timber, reduced dead weight and ease and speed of erection.
- Improved thermal insulation reducing life cycle costs.
- Simplified details in terms of embellishment and style.
- The overall lighter roof enables a more slender construction to be employed.
- The introduction of timber impregnation reduces the possibility of future decay.
- The longer spans of roof trusses helps to eliminate internal load-bearing walls.
- The reduced floor to ceiling heights reduces both the initial cost and costs in use.

The above aspects do not reduce building performance but often enhance it, and at a reduced cost, thereby adding value for the owner and user.

3.6 Procurement

Traditionally, a client who wished to have a building constructed would invariably commission an architect to prepare drawings of the proposed scheme and, if the scheme was sufficiently large, employ a quantity surveyor to prepare appropriate documentation on which the building contractor could prepare a price. These would all be based upon the client's brief, and the information used as a basis for competitive tendering. This was the common system in use at the turn of the century and still continues to be widely used in practice.

Particularly since the mid-1960s a small revolution has occurred in the way designers and builders are employed for the construction of buildings. To some extent these are the results of initiatives taken by the then Ministry of Works in the early 1960s and the Banwell (1967) committee. This recommended several changes in the way that projects and contracts were organised, one of which was an attempt to try to bring the designers and the constructors closer together. Much later the Latham (1994) report considered how such evolving arrangements and general organisation might be improved.

There is, however, no panacea, since each individual project generates its own peculiar characteristics. Even projects that appear to be identical result in different outcomes in terms of cost, time and quality. However, projects with similar characteristics may be grouped together for procurement purposes. The over-riding characteristics are those that emulate directly from the client, and are set out in the brief, often in terms of aims and objectives. Different clients will define these differently, and some clients the lowest possible price set against clear definitions of quality will always remain a prime consideration. Buying cheaply may not be good advice, but purchasing buildings against prescribed criteria at the lowest price is what most of us expect to achieve in our various walks of life. Building costs will always remain an important consideration as clients seek to secure added value from their projects. For other clients, time will remain paramount, but few clients will pursue this at the total disregard of cost. In extreme situations, where the project will generate considerable income when in use, then building costs may diminish in terms of their significance. Emergency works projects may also place cost near to the bottom of their objectives. These do not represent the typical or even common sort of project. Quality in buildings has now almost become a byword, if only to respond to the poor quality that has been so much in evidence during the latter part of the twentieth century.

Each of the methods available for the procurement of buildings (Ashworth, 1996) has its own particular characteristics, advantages and disadvantages. All have been used in practice at some time. Some have become more popular than others largely due to their familiarity and the consequent advantage of ease of application. New methods will continually evolve to meet the rapidly changing circumstances of the construction industry. These will occur in response to current deficiencies and to changes in the culture of the construction industry, and the key issues are:

- Consultants v. Contractors
- Competition v. Negotiation
- Measurement v. Reimbursement
- Traditional v. Alternatives

Procurement arrangements are dependent on the following:

- **Type of work to be performed**: Building, civil engineering, process plant engineering.
- **Size of project**: Forms are available for major and minor works and those of an in-between nature.
- **Status of designer**: Architects are more likely to prefer JCT or ACA, whereas civil engineers will opt for an ICE or NEC form.

Figure 3.2▸ Clients' requirements

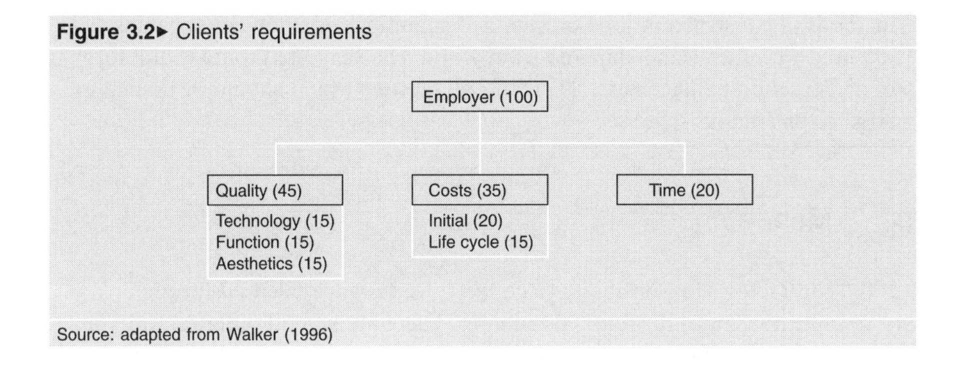

Source: adapted from Walker (1996)

- **Public or private sector**: Different forms are available for use by private clients and local and central government. Large industrial corporations may in addition have their own forms of contract.
- **Procurement method to be used**.

Changes in emphasis that are believed to add value include: single-point responsibility, early involvement of the contractor in the project, team working, correct allocation of risk, and trust. The employer's main requirements continue to be:

- The layout for the project should be acceptable in terms of function and use.
- The design should be aesthetically pleasing, not only to the employer but also as viewed by others.
- The final cost of the building should closely resemble the estimate.
- The quality should be in line with current expectations.
- The future performance of the building and the associated costs-in-use should fit within specified criteria.
- The project should be available for hand-over and occupation on the date specified for completion of the works.

These are set against criteria that can be broadly summarised in Figure 3.2. The different weightings will vary directly with the aims and objectives that are applied by different employers.

Latham (1994) identified three different scenarios.

- **Traditional construction**: These are the projects that use the normal techniques of design and construction. Standard forms of contract are used, and this is the route with which the industry is most familiar. Problems do occur through lack of co-ordination between design and construction but these can be minimised through effective project administration.
- **Standard construction**: This is the type of project where the end product can be achieved through a predetermined construction route, probably involving a limited range of standardised processes and components. This arrangement is perhaps best served by a design-and-build arrangement.
- **Innovative construction**: In these circumstances the client commissions a project which involves a high degree of innovation in terms of its design and the

methods of construction used on site. A demanding procurement system is required with firm leadership and team work. The suggested route is of a form of construction management. There are no reasons why it should be limited to large or prestigious schemes.

3.7 Management

It is the contractor's responsibility to convert the design, evidenced through plans and specifications, into the finished building. There are several different ways in which this can be achieved, each offering its own advantages and disadvantages. As with procurement there is no panacea. Designing, costing, forecasting, planning, organising, motivating, controlling and co-ordinating are some of the roles and responsibilities of those involved in managing the design through to completion (Construction Industry Training Board, 1991).

The construction industry is characterised by a large number of small firms (Harvey and Ashworth, 1997) and a relatively small number of large contractors that carry out the industry's workload. In 1999, about one-third of the workload was completed by the eleven largest firms. In practice, however, much of this work is subcontracted to specialist or smaller companies. Consequently, the planning process follows a very similar pattern. According to Cooke and Williams (1998) the planning process may be considered in three distinct stages:

- **Pre-tender planning**: Planning during the preparation of the estimate and prior to submitting the final bid to the client.
- **Pre-contract planning**: Planning prior to commencing the work on site.
- **Contract planning**: Planning during the construction process, which also includes procedures to ensure that the plan is being achieved.

Several different techniques have been devised to attempt to ensure that the complexities involved in modern building will still allow the plan to be completed on time, at the agreed cost and at the defined quality.

3.8 Facilities management

The many and varied functions that are involved in facilities management are not new but the way that these are managed has become more integrated and extended. The role of a facilities manager is recognised as being more proactive and different to that of the traditional estates manager. Facilities management also differs from property management in that the range of activities that it covers far exceeds matters other than just buildings and property. Facilities management is concerned not just with the building structure, services and finishings but with the activities that go on within the building. It is much more than a combination of the traditional disciplines of estate management and building maintenance. It includes a range of activities that can be broadly summarised under the seven headings given below (source: *Premises and Facilities Management Journal*). This division of activities will vary depending upon the type, nature and size of organisation concerned.

- Building maintenance and operations:
  - Electrical
  - Mechanical
  - Fabric
- Environmental management:
  - Energy management
  - Health and safety
  - Waste management
- Infrastructure:
  - Utilities
- IT and telecommunications:
  - Telecommunications services
  - Systems administration and management
  - Customer response support
- Property management:
  - Asset management
  - Space planning
- Support services:
  - Catering and vending
  - Cleaning
  - Office support services
  - Security
- Transport:
  - Fleet management

The professional role of facilities managers has been increasingly recognised over the past few years. Their emphasis has been on improving added value in the provision of traditional facilities services. The use of external consultants for this purpose has largely been the result of their promise to reduce costs while maintaining or improving quality. This has been achieved through the wider experience gained by consultants in managing different clients and by having a clearer focus on their work, not being distracted by in-house politics. Consultants are better able to disseminate good practices because of their contacts with a range of different clients. External consultants also claim to remove some of the hassle from the client. The use of external consultants is partially threatened by the demands of accelerating organisational change. New work practices, such as hot desking, hoteling and remote working call for changes in the way that the workplace is being used.

Facilities management has grown out of a range of different disciplines, most of which came from general management, as shown in the above list. About one-third of practising facilities managers are members of the British Institute of Facilities Management (BIFM). As many as 65% hold other professional qualifications but less than 10% have any academic qualifications in facilities management. Qualifications and skills development are therefore two important priorities for facilities managers. The BIFM's national professional qualification and training programme is being used by major public and private sector organisations.

3.9 Quality assurance

Every employer of construction firms in the construction industry has the right to assume a standard of quality that has been specified for the project. The Building Research Establishment (BRE) has, in conjunction with a number of sponsors, launched the Construction Quality Forum almost ten years ago. At the launch of this organisation, BRE reported that each year defects or failures in design and construction cost the industry, and its clients, more than £1 billion. This represents 2% of total turnover, which is a high percentage taking into account that contractors typically make only 6% profit on turnover (Harvey and Ashworth, 1997).

Construction is an industry in which:

- There has never been a requirement for the workforce to be formally qualified and skills are generally developed through time serving.
- Much of the work is carried out by subcontractors in a climate in which some 50 firms come into existence every day and a similar number go into liquidation or become bankrupt every day.
- There is a paucity of research, development and innovation involving new materials and designs.
- There is often poor management and supervision.

Studies have indicated that about 50% of faults originate in the design office, 30% on site and about 20% in the manufacture of materials and components. An investment in quality assurance methods can therefore reap substantial long-term benefits by helping to reduce such faults, the inevitable delays and costs of repairs, and the all too frequent legal costs that often follow.

ISO 9000 certification has been increasingly taken up within the construction industry by consultants and contractors. Quality assurance is therefore seen as a good thing for the industry. Contract procurement methods that fail to address this issue adequately are not doing the industry or its employers any favours. The use of quality certified firms, who have been independently assessed and registered, therefore offers some protection within the context of getting it right first time. Work that is below acceptable quality and standards, and has to be rectified, is rarely as good as ensuring that it is carried out correctly in the first place.

ISO 9000 is seen by some firms as an additional expense, with the cost of accreditation as an unnecessary overhead. Also, while it should ensure that quality standards are achieved, it does not ensure that the appropriate quality has already been set in the first place. However, some employers now prefer to employ consultants, contractors or suppliers who have a kitemark of good quality work. Within the total quality management scenario, quality remains an ongoing process of continued improvement. Quality must be appropriate to the work being performed, and exceptional quality should only be insisted upon where it adds value to the finished construction project.

A major comment of Latham (1994) was that quality should be the over-riding consideration.

3.10 Adding value

The thrust of changes to design and construction, and the utilisation of new technology, have been aimed at adding value to a client's project. This can be seen in the way that the project is designed in respect of its functional performance. Many buildings become obsolete because the purpose for which they were originally designed has now changed or evolved to such an extent to make the continued use of such a building difficult. This can best be seen in respect of industrial buildings for manufacturing purposes as some manufacturing processes have changed to such an extent that the existing buildings will no longer meet their new function. Domestic buildings can also be affected in the way in which the space is organised. If, for example, more people choose to work from home rather than in the office, then this will have implications for new houses. This phenomenon is exactly the opposite to that which affected weaver's cottages in the nineteenth century.

New technology, as a result of research and development, has also enabled us to use different methods for construction purposes. This allows the constructor to complete the project more quickly than previously and sometimes to more exacting standards. These add value to the project. Advances in information technology have enabled constructors to utilise this to carry out routine operations more effectively and more precisely. Machine-made components can be made to more exacting and repetitive standards.

The use of much modern technology is also incorporated into the finished project. Perhaps one of the greatest changes in modern buildings is the inclusion of engineering services and automatic engineering control systems that are now provided on all projects. These add value to the project while in use by the occupier and have become an essential facet, required on all but the simplest of projects. Comparison of the escalation of the use of these items and their respective costs can easily be demonstrated, even from the mid-point of the twentieth century.

References and bibliography

Ashworth A. (1996) *Contractual Procedures in the Construction Industry*, Longman.

Banwell P. (1967) *The Placing and Management of Contracts for Building and Civil Engineering*, HMSO.

Brown M. A. (1989) *The Application of Robotics and Advanced Automation to the Construction Industry*, The Chartered Institute of Building, Occasional Paper No. 38.

Construction Industry Training Board (1991) *The Construction Industry Handbook*, Hobsons Publishing.

Cooke B. and Williams P. (1998) *Construction Planning, Programming and Control*, Macmillan, London.

Fletcher Bannister, Sir (1996) *A History of Architecture*, Architectural Press.
Harvey R. C. and Ashworth A. (1997) *The Construction Industry of Great Britain*, Butterworth-Heinemann.
Langsten C. (1997) *Sustainable Practices: ESD and the Construction Industry*, Envirobook Publishing.
Latham M. (1994) *Constructing the Team*, HMSO.
Moore J. F. (1992) *Monitoring Building Structures*, Blackie.
Premises and Facilities Management Survey (1998) *Premises and Facilities Management Journal*, March.
RICS (1990) *Planned Building maintenance: A Guidance Note*, Surveyors' Publications.
Walker A. (1996) *Project Management in Construction*, Blackwell Science.

Chapter 4

Cost planning

4.1 Introduction

Before the early part of the nineteenth century, the costs of building were established using a principle of *measure and value*. When building works were completed they were measured and the quantities were valued in accordance with a schedule of agreed rates. The costs of building were not established until the project was completed. The system was largely replaced by bills of quantities for competitive tendering purposes as early as the 1840s. The coherence was hastened by the replacement of the craft system and the introduction of general contractors. The main form of early cost advice and early price estimating at that time was based on the cube rules and the superficial floor area method.

Following the end of the Second World War and a period of austerity a massive reconstruction took place throughout many towns and cities. This redevelopment included hospitals, schools and housing. For a time there was a restriction on what could be built due to the lack of availability of materials and a shortage of skilled workers. Private projects, therefore, took longer to develop before reconstruction could commence. The building boom, because this is exactly what this period of history was, was fuelled by politicians on the strength that many public buildings had been destroyed by war. Repair and maintenance during 1939–45 had been restricted to essential work alone. In fact only limited repair and maintenance work had been carried out throughout the period since the end of the First World War and the economic depression of the 1930s. The importance that is given to these activities had not yet recognised the importance of maintaining capital assets other than on a 'make do and mend' philosophy. Coupled with these two problems were the aspirations given to post-war children of great expectations. Better housing, improved medical care and longer periods of education also placed requirements on the construction industry to provide '*homes fit for heroes*'. Coupled with these was the demand for improved infrastructures and associated engineering works. The advent of extensive motor car ownership and the need for new roads and motorways were still being considered. The first stretch of motorway in Britain, known as the Preston bypass, was not opened until 1959. In order to meet these many and varied demands the public sector introduced the principles of cost limits for capital works projects (Spedding, 1982).

Three important factors were instrumental in cost control and cost planning methods:

- The need for the large-scale public sector building programme referred to above.
- The trend towards using non-traditional building materials and industrialised systems of building, increasing the complexities of design.
- The increasing cost awareness of clients in both the public and private sectors who required improved cost advice and a cost forecast prior to committing their finances to a project.

4.2 Cost planning

'Cost planning' is a term that can be used to describe any system of bringing cost advice to bear upon the design process. In the context of the development of construction economics the term has developed its own significant meaning, describing a process that has evolved since its first inception. Cost planning, to be effective, requires a close working relationship and co-operation between engineer, architect and quantity surveyor. It also requires an appreciation of each other's objectives.

Cost planning – an economic technique that was also applied to buildings – was introduced barely fifty years ago. There has never been any question that building should not be commenced until the design has been properly planned and the method and plan for the construction process have been determined. Within this there is a framework that controls planning on a macro scale through town planning and Building Regulations approval. In today's society it is considered foolish to proceed with any event without properly considering, as far as possible, all of the consequences involved.

In order to obtain the greatest number and quality of buildings for the limited finances that were available, new methods had to be developed. The single price methods that were used for estimating the approximate costs of projects were too blunt a tool to provide a method of planning expenditure under a number of different budget headings. While the use of building trades were considered for this purpose, they were of only limited use and bore no relationship for the comparison between different projects. They thus provided little information that would aid the planning of the costs of buildings. Also designers (architects) did not design projects on a trade by trade basis. Their designs frequently evolved from outline sketch schemes by considering the function of different parts of the building, i.e. building elements. Since buildings were frequently design-led fifty years ago, it was a sensible approach to develop a cost planning system that recognised this fact and assisted the designer during the process of design.

Other methods of cost planning and analysis were considered, such as the percentage allocation method. This assumed that specific building elements of similar buildings were proportional in their costs. In projects other than those of a repetitive nature this assumption is not well founded.

The essence of cost planning is to allow the architect to control the cost of a project, i.e. to keep it within its target cost while it is still being designed. The earlier this process is introduced, the greater is the measure of control that can be excerised over the ultimate cost, quality and design. Cost planning is a continuous process with progressive checks being made on the costs involved in the individual elements as the design develops.

Modern cost planning should consider all the client's costs associated with an individual construction project. These will include the initial construction costs and the costs in use that make the project effective throughout its life cycle. It will also include a consideration of grant availability and taxation requirements, together with the costs of finance. These assist in assessing the financial effectiveness of the project as a whole.

4.3 Principles of cost planning

The objectives of cost planning are aimed at:

- Achieving value for money for a client.
- Sensible expenditure between the different parts of a building.
- Keeping the total expenditure within the amount agreed by the client.

The underlying principles involved have been defined as follows (RICS, 1976):

- There is a standard reference point for each definable part of a building.
- It allows performance characteristics to be related to each reference.
- It allocates costs in an apportioned and balanced way throughout the building.
- Previous projects can be classified in a standard manner.
- The same process and procedures with design methods are adopted.
- Costs can be checked and amended as the design develops.
- Designers can take the necessary action before committing themselves to any one design solution.
- Design risks and contingencies can be taken into account.
- Costs can be presented in a logical and orderly way to the client as the design develops.

In projects that are planned cost effectively, the standard procedure will often be activated by a proposal to consider several different options before making a design decision. Thus a designer may wish to know the relative or absolute costs of different types or configurations of structural frame. Equally the cost implications of alternative procurement arrangements will need to be considered. The outcome of such exercises will be a statement of the relative costs of the options together with an assessment of their acceptability within the budget. It is important that all cost advice is related to the agreed budget, since cost planning assumes that the clients want to control their financial commitments (Bennett, 1981).

4.4 Cost-planning systems

Two separate systems of cost planning evolved, known as *elemental* and *comparative* cost planning.

In 1953, the then Ministry of Education invented cost planning. This was described in *Building Bulletin No. 4* (1957). It provided an elementary description of the proposed method to be used. *Building Bulletin No. 4* devoted its entire contents to 'cost study' and introduced the industry to the principles involved in cost planning. Between 1953 and 1957 there had been substantial developments and experience in its use. An increasing number of architects and quantity surveyors were finding it a powerful instrument in the control of building costs. The British Architects Conference in 1956 organised by the Royal Institute of British Architects (RIBA) on the subject of architectural economics devoted a full and detailed discussion to the subjects of cost analysis and cost planning.

The interest in this technique at the time was two-fold. Building costs and prices were rising rapidly during the post-war years and thus, for the total amount of budget that was available, fewer school buildings could be constructed. In 1956, Circular 301 introduced the concept of building cost limits, which restricted the amount that could be spent on buildings. This limited initial building costs to a ratio of the number of pupils for whom the school was being constructed. The existing methods that were used at the time to control building costs were largely ineffective at meeting these goals. The introduction of cost planning into the field of building costs was a development of profound importance to building and architectural economics. Cost planning properly covered the interests of four parties in the building process: client, architect, quantity surveyor and building contractor.

In 1956, the Royal Institution of Chartered Surveyors set up its Cost Research Panel, which was later instrumental in developing the Building Cost Advisory Service which would eventually be renamed the Building Cost Information Service. About the same time the Royal Institute of British Architects also had its own Cost Research Committee.

Elemental cost planning

This method is sometimes referred to as target cost planning since a cost limit is fixed by the financial method of approximate estimating (see Ashworth, 1999), and the architect must then design the project to fit within this cost. The cost plan is the architect's design in financial terms. The process is therefore described as *designing to cost*. The method has been used extensively throughout the public sector, where financial limits were placed on proposed expenditure.

Comparative cost planning

This method of cost planning is often described as *costing a design*. The main difference between this and elemental cost planning is that no fixed element budgets are necessarily implied. Different design solutions for each element are considered on their respective merits, with the design team and the client making the appropriate selection based upon a range of design and cost information. The most expensive solution may be selected, but this decision is taken in the full knowledge and awareness of the cost consequences.

In practice, the cost-planning process is usually a combination of the two theories. Most projects recognise the need to set element cost limits, but not necessarily at the strict expense of eliminating the advantages of an overall good design solution. This overall aspect should include the full consideration of life cycle costs, which could, as far as building costs are concerned, increase initial costs. This fits easily within the concept of not necessarily choosing the least expensive design solution, but of balancing design benefits with their cost implications.

Advantages of cost planning

The general advantages claimed for the use of cost planning include:

- The tender sum is more likely to equate with the approximate estimate.
- There is less of a possibility of abortive redesign being necessary at the tender stage due to tender sums being higher than expected.
- Some element of cost effectiveness through a balanced design is more likely to be achieved.
- Cost considerations are more likely to be considered as a full and integral part of the process.
- The amount of pre-tender analysis and evaluation by the design team should result in a smoother running of the project on site.
- Cost planning provides a useful means of comparing the costs of individual projects.

As with all advances and changes in practice, some disadvantages are likely to emerge. The designer, for example, may feel that cost planning impinges too much on the design and restricts the thought process of the designer. The system is also more time consuming in preparation, reporting and consideration of the cost plans. Additional time and the subsequent fees involved are therefore required.

4.5 The cost-planning process

While there are two systems that are used for cost planning, the processes involved are similar and include the stages shown in Figure 4.1. Figure 4.2 presents an adaptation of the cost-planning and design process given in *Building Bulletin No. 4*.

Early cost advice

This is usually done to provide a client with some indications of the likely costs involved based upon a client's broad aims and objectives regarding the project. The methods used for financing the project and the determination of its general financial viability would also be provided at this stage.

Figure 4.1▸ The cost-planning process

Early cost advice	A discussion with the client on the broad aims and objectives of the project.
Preliminary estimate	This is prepared usually using a single price method of approximate estimating.
Preliminary cost plan	The preliminary allocation of the budget among the various elements of the project.
The cost plan	This is based upon an elemental analysis involving the setting of cost targets for the whole project.
Cost checking	The evaluation of changes in the design of elements during the design process.
Tender reconciliation	The comparison of the final cost plan with the accepted tender sum.
Post-contract cost control	The control of costs during the construction period.

Figure 4.2▸ Cost planning and the design process

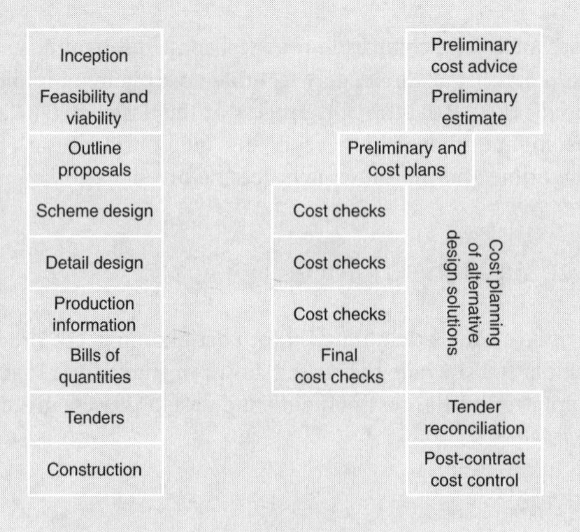

Source: adapted from *Building Bulletin No. 4*: Ministry of Education (1957)

Preliminary estimate

This is usually prepared at the outset of a project when a quantity surveyor is asked to provide some ideas on the likely project cost. If drawings are available these are likely to be sketches that may provide an indication of the size (floor area), possibly the materials that might be considered and ideas of how the client's brief may be developed. If the client undertakes a considerable amount of building work, then the information and the brief are likely to be more precise but otherwise many design decisions will still need to be resolved. With the more informed clients some idea of the likely expenditure that is involved will also be known. The preliminary estimate attempts to confirm, or otherwise, the client's overall expectations. If the project is a one-off scheme for the client, then the information is likely to be less precise with aims and objectives that will need to be refined as the project develops.

The methods used for the preparation of early price estimates have been described in Ashworth (1999). Many of these are applied in a deterministic way providing clients with a single estimate of total cost that is used for budget purposes. The assessment of design and price risks in preliminary estimates are also considered in Ashworth (1999). This indicates that the risks associated with each of these factors are greater at inception, where many decisions still have to be taken, than at later stages when the project is more fully developed and where costs have become more explicit. It is now generally considered to be insufficient to provide clients with only a single estimate of cost; a more appropriate way is to offer a range of possible values for the estimated tender sum. The technique of multiple estimating using risk analysis (MERA) attempts to provide such a range of estimates. This procedure was devised by the former Property Services Agency within the Department of the Environment.

Traditionally, early price or approximate estimates are prepared to provide clients with a budget of their expected costs. This is to avoid the expensive design fees on aborted schemes that have to be abandoned because they are too expensive. Historically, no measure of accuracy or confidence limits were given; the estimate would have been prepared in accordance with the reasonable skills and expertise that was expected. Estimate deviation between preliminary estimates and contractors' tenders were usually explained away, with suitable explanations for any discrepancy being provided.

It is considered to be good practice today to provide preliminary estimates of cost within a range of values and to offer confidence limits on these values. Such information is now automatically requested by informed clients and this provides them with additional information on which to make sounder judgements.

Providing the preliminary or initial cost estimate is therefore crucial to the whole scheme, and is usually concerned with the seven factors listed below. This requires considerable skill, experience and judgement to be performed by the person preparing the preliminary estimate. Prior to cost planning being introduced, this is where pre-design estimating ceased. The next cost information to be provided was supplied by the contractors' tenders sums.

- **Market and contract conditions**: Factors to consider – workloads, labour availability, type of client, contractual conditions, etc.

- **Design economics**: The effects on cost of the design, shape, height, size, constructional details, etc.
- **Quality considerations**: The quality of materials and the standards of workmanship, including the compliance with government regulations.
- **Engineering services**: The type and amount of engineering services.
- **External works**: The size of the site and the features that are to be provided.
- **Exclusions**: Items excluded from the estimate would include not only land costs, VAT, professional fees, interest charges and loose furniture, but also such items as the fitting out on some projects.
- **Price and design risk**: Allowances to cover design and its impact on construction methods and the volatility or otherwise of the construction market.

Preliminary cost plan

The preliminary cost plan is really the first phase of the cost-planning process. Its main purpose is to determine the cost targets for the different elements involved. The preliminary estimate will have been accepted by the client and the architect. This estimate may be subject to revision based upon a preliminary investigation of the site and further consideration of the proposals outlined in the client's brief. A sum will eventually be accepted that satisfies the brief and the proposed design, and this will form the basis of the cost target. Alternatively, there will be circumstances in which the target expenditure is imposed, as in the case of government-sponsored projects. The evidence in these cases suggest that, for defined standards of specification and the amount of accommodation required, it is possible to design within a cost limit.

Where the proposed scheme is similar to that of another project that has already been constructed and previous analysis is available, the previous project can be used as a basis, but will need to be adjusted for inflation and regional differences in costs. In these circumstances, where the target cost will have been derived from other similar buildings in terms of standards and quality, it may be necessary to make adjustments for site conditions, the construction technology used or architectural requirements. Alternative methods of construction may be evaluated and aspects of contractual implications considered. Where important design decisions still have to be taken, perhaps in relation to height or shape, the preliminary cost plan will need to be delayed until such important design decisions can be made. There is little point in attempting to cost plan a project solely on the basis of total floor area, since building morphology can have an important influence on costs.

At the early stages of projects the range of possible solutions is considerable. The range of costs associated with these solutions is also very wide. An object of cost planning is to guide the design and organisation decisions, to ensure that the answer that is finally accepted is efficient and is able to be provided within the budget (Bennett, 1981).

The cost plan

Once the sketch plans have been completed and agreed by the client, the task of allocating sums of money to the various elements can take place. The individual element unit target costs must of course equate with the total target cost, otherwise the cost plan will be out of balance. It may also be desirable to present these within a range of probable costs, although it is essential that an overall agreed total cost is also identified. The methods described for cost analysis will be employed to arrive at element unit totals. The following information will be required:

- **Drawings**: at least plans and elevations.
- **Specification**: an indication of the *quality* of materials and *standards* of workmanship.
- **Contracts**: the likely method of appointing a contractor.
- **Cost analysis**: available from other comparable projects.
- **Other analyses**: to be used as a second opinion.

The calculated target costs should be adhered to by those involved in the project. Where any differences are agreed these should be recorded for information and appropriate action. A separate sum should be set aside for price and design risks. This amount will vary depending upon the experience and skills of the designer and quantity surveyor and the nature of the client. The percentage adjustment can vary considerably depending upon the complexity and innovation in the design. The adjustment will be reduced the closer the project gets towards the final design phase and the tender date.

In preparing a cost plan the following need to be considered:

Pricing conditions

In the absence of local information, building cost and tender price indices can be used to indicate how the relative price of building is changing. These indices help to explain inflation and changes in market conditions. The cost planner is faced with extrapolating this information to cover the proposed project.

Quantity considerations

There are essentially three ways in which adjustments for quantity can be made during cost planning:

- **Approximate quantities**: This is the most popular way and involves a simplified measurement of the elements in accordance with the rules that have been adopted for pricing purposes.
- **Proportion**: An existing cost analysis can be used as a basis for calculating the elemental costs involved. This is the preferred method since the cost plan will then automatically include the requisite allowances for all of the elements in the proposed building. It necessitates the adoption of quantity factors.
- **Inspection**: Some elements are difficult to quantify realistically at the cost-planning stage and their values can only be assessed subjectively.

Quality considerations

Quality is difficult to quantify and hence to value. Qualitative adjustments can be made where more expensive materials are to be included within a project. The client and the designer are the best indicators of quality, although sometimes what is expected is not always specified.

Cost checking

During the design stage of the project the different components or elements of the design evolve. Throughout this process it is necessary to compare the costs in the evolution of design with the amounts allocated in the cost plan. Designers usually design in this way and cost planning, in adopting the elemental approach, allows the changes in costs to be easily calculated and compared. Where the architect does not diverge from the information and ideas incorporated in the cost plan, the time and effort involved in cost checking will be minimised. This is frequently achieved in those circumstances where similar projects are being designed for the same client. In the case of an individual design to meet a one-off solution great care will be required if the desired results of cost planning are to be achieved.

Where cost checking reveals compatibility with the original cost plan, then only limited action will be required. It must be appreciated that an element may be designed and redesigned several times in order to achieve the correct solution. In a similar way cost checking will also replicate this process. Where a cost check reveals differences from the target cost, then different courses of action are required, as follows:

- Redesign the element so that the target cost for the whole scheme is not affected.
- Approve the change in the element unit total and amend the overall cost plan amount.
- Approve the element unit total but examine ways in which other element unit targets can be amended so as to leave the cost plan total unchanged.

It is always dangerous to assume that reductions in the costs of future elements, yet to be designed, can be achieved. This should never be assumed. The element unit totals should not be adjusted unless there is good evidence to support the case.

The cost checking should be carried out as soon as the design details are received, in order to attempt to reduce any future design work, should the element total be too costly. The advantages to clients of this process is to provide them with the full implication of design decisions that they approve. It provides a reasonable assurance that the budget estimate will not be exceeded. The greatest amount of attention by the quantity surveyor should be given towards the cost-sensitive elements, but this does not imply that other less important elements should be ignored. The pricing of cost checks will be carried out using current market prices. During the later stages of the process, the costs of specialist works will be replaced with firm quotations from designated suppliers and subcontractors.

Cost checks are sometimes made on the visual inspection of the design details, and great care is required to ensure that items are not overlooked. Where there are no changes to the cost of an element between present and past design solutions, a nil return should be reported.

The usual method of cost checking involves the use of approximate quantities and current all-in prices. A final cost check is usually made when the scheme is with contractors during their own tendering process. Where bills of quantities have been prepared, these can be priced prior to receipt of tenders.

Tender reconciliation

Where the process of cost checking has been carried out thoroughly, the receipt of tenders should provide few surprises. It is nevertheless appropriate to carry out certain checks on the tender to be accepted, to highlight any differences between this and the final cost check. This will provide some insight into the cost planning of any future projects. Any discrepancies between the two should then be easily explained.

Errors on the part of the quantity surveyor and the builder's estimator do occur. The deliberate distortion of individual tender prices by the contractor may be made for possible future gain. The quantity surveyor will report on the sufficiency of the contractor's prices which will include both a technical and arithmetic check of the tender sum that is recommended for acceptance.

Post-contract cost control

The cost-planning process does not cease when the contractor starts work on site. The design is likely to have stayed within budget, evidenced by the accepted tender. While the design may change little, there are provisions for variations authorised by the architect or other designers. Some of these changes will be due partly to site conditions, revisions instigated by the client and changes made by the designer. The client's main financial interest will now be to ensure that the budget and final cost remain in agreement. While the budget and tender cost will have broadly achieved this, changes and unforeseen items of expenditure can result in unacceptable differences.

Cost planning during construction is normally achieved through the regular preparation of financial statements. These will advise the client of any probable changes between the agreed tender sum and the probable final cost of the project. The size and complexity of the project will determine how frequently these statements are prepared. The financial report is in two stages. The first considers the current financial position of the project and the second the expected final cost based upon adjustments to the contract sum. These may include the costs of any variations, adjustments to prime cost and provisional sums, daywork accounts, increased costs and the possibility of contractual claims.

4.6 The effects of cost planning on costs in use

Cost planning that only considers the initial costs of construction without considering the life cycle cost aspects is limited. It is recognised today (see Chapter 5) that a correct balance and consideration of life cycle costs is required to provide full benefits for the client. It may be sensible, finances permitting, to spend a small amount extra on initial costs thereby reducing the costs in use throughout the project's life. However, it cannot be assumed that increased expenditure today will necessarily reduce the overall life cycle costs. In fact it can easily be demonstrated that spending more on initial costs will have the effect of increasing the life cycle costs of a project. Increasing expenditure on the size and scope of the project, installing additional engineering services or, in some cases, providing a higher quality will have this effect. In the example of the latter item, building components with low initial costs can often cost higher amounts for their maintenance, but an increased initial cost, properly expended, can have the effect of reducing overall life cycle costs. The whole premise of life cycle costs is based upon this general, albeit not universal, assumption. There needs to be an appropriate balance between initial costs and costs in use.

The cost-planning process must therefore carefully consider the implications of initial design proposals on life cycle costs. The adoption of a cost limit approach to individual elements must therefore fully consider the maintenance standards to be achieved in practice. The situation is further complicated when taxation and associated capital allowances are taken into account. It is often argued that since the majority of capital building costs are not allowable against taxation (see Chapter 11), whereas some maintenance costs are, then there is little point in attempting to reduce the latter if more is required to be spent on capital costs. There is the need to place a greater emphasis on the total cost evaluation of construction projects, aiming to strike the correct balance in the initial design solution.

4.7 Conclusions

There can be little doubt that planning before action is of great benefit in all walks of life. Cost planning is a process that plans out the expenditure on a building prior to committing a client's finances to a project. It adds value for the client. It is difficult to see how effective a project can be in terms of its forecasted costs without this consideration. Lump sum estimates, or only determining the costs after completion, are inappropriate techniques for the present age. The use of cost planning also allows costs in use to be more carefully considered and as such, therefore, provides a foundation for the application of all other costs. The application of cost planning further adds value to the client by allowing each element to be considered not just in terms of its initial cost but in the wider context of life cycle costing.

References and bibliography

Ashworth A. (1999) *Cost Studies of Buildings*, Addison Wesley Longman.

Bennett J. (1981) *Cost Planning and Computers*, Property Services Agency within the Department of the Environment.

Ministry of Education (1957, revised 1972) *Building Bulletin No. 4: Cost Study*, HMSO.

RICS (1976) *An Introduction to Cost Planning*, Royal Institution of Chartered Surveyors.

Spedding A. (1982) What happened to the school building cost limit? In P. S. Brandon (ed.) *Building Cost Techniques: New Directions*, E. & F. N. Spon.

Chapter 5

Life cycle costing

5.1 Introduction

If we consider the world as a whole, it is clear that the resources available are limited and insufficient to meet all of the wants of mankind. With this in mind it has long been considered unsatisfactory to evaluate the costs of buildings and other structures on the basis of their initial costs alone. The only fully comprehensive view of construction costs is to consider all of the costs associated with an investment. These include both present and future costs over the entire life cycle of the project. Different clients will have different aims and objectives, and costs will be considered in different ways, but a sensible client will not choose to ignore them.

Life cycle costing involves the application of established economic techniques to the decision-making process associated with the design and commissioning of capital works projects. The combination of initial capital expenditure and future costs-in-use may be fitted to the constraints of the client and the project under consideration. It is a trivially obvious idea that all costs arising from an investment are relevant to that decision. By definition the image of a life cycle is one of progression through a number of different phases.

The pursuit of economic life cycle costs is the central theme of the whole evaluation. The method of application incorporates the combination of managerial, functional and technical skills in all phases of the life cycle. The proper consideration of the costs-in-use of a project is likely to achieve improved value for money and improved client satisfaction. Different objectives may be set at the different times of the project's life.

5.2 The importance of long-term forecasting

Forecasting is required for a variety of purposes such as early price estimating, the setting of budgets, invitation of tenders, cash flow analysis, final account predictions, and life cycle costing. While it is recognised that there are confidence and reliability problems associated with initial cost estimating, these are of a smaller magnitude than those associated with life cycle costing. A large amount of research has also been undertaken in an attempt to improve the forecasting reliability of early cost advice. By comparison, the acquisition of life cycle costing knowledge and skills through research and application are still in their infancy, with a considerable gap between theory and practice. It is also difficult to provide confidence criteria, due largely to an absence of historical perspectives, professional judgement and a feeling for a correct solution. The fundamental problem associated with the application of life cycle costing in practice is the requirement to be able to forecast a long way ahead in time. While this is not in absolute terms, it must be done with sufficient reliability to allow the selection of project options which offer the lowest whole life economic solutions. A major difficulty is therefore related to predicting the behaviour of future events. While some of these events can at least be considered, analysed and evaluated, there are other aspects that cannot even be imagined today. These remain therefore outside of the scope of prediction and probability, and are thus unable to be even considered, let alone assessed in the analysis. The key criteria, however, for life cycle costing is not so much in the accuracy of the forecast, but in allowing the correct economic solution to be made.

5.3 Life cycle costing applications

Life cycle costing is not solely a project design tool; it has applications throughout the project's entire life (Ashworth, 1993).

At inception

Life cycle costing can be used as a component part of an investment appraisal. This is the systematic approach to capital investment decisions regarding proposed projects. The technique is used to balance the associated costs of construction and maintenance with rental values and needs expectancies. It is a necessary part of property portfolio management. It recognises that many projects are built for investment purposes. The way that future costs-in-use are dealt with therefore largely depend upon the expected ownership criteria of occupation, lease or sale, or indeed a combination of these alternatives.

At the design stage

A main use of life cycle costing is at the design stage or pre-contract phase of a project. Life cycle costing can be used to evaluate the various options in the design in order to assess their economic impact throughout the project's life. It is unrealistic to attempt to assess all of the items concerned, indeed the cost of undertaking such an exercise might well rule out any possible overall cost savings. The sensible approach is to target those areas where financial benefits can be more easily achieved. As familiarity with the technique increases, it becomes easier to carry out the analysis, and this may prompt a more in-depth study of other components or elements of construction. While some of the areas of importance will occur on every project, others will depend upon the type of project being planned. For example, roofing is probably an important area for life cycle costing on most projects, whereas drainage work is not. However, on a major highway scheme, where repeatability in the design of the drainage work occurs, the small savings that might be achieved through life cycle costing can be magnified to such an extent as to make the analysis worth while. The important criteria to adopt is that of cost sensitivity in respect of the whole project costs.

Life cycle costing is perhaps most effective at this stage in terms of the overall cost consequences of construction. It can be particularly effective at conceptual and preliminary design stage, where changes are able to be made more easily, and the resistance to such changes are less likely, than when a design is nearing completion. In these circumstances the designer may be reluctant to redesign part of the project even though long-term cost savings can be realised.

At procurement

The concept of the lowest tender bid price should be modified in the context of life cycle costing. Under the present contractual and procurement arrangements manufacturers and suppliers are encouraged to supply goods, materials and components which ensure their lowest initial cost irrespective of their future costs-in-use. In order to operate a life cycle cost programme in the procurement of capital works projects, greater emphasis should be placed upon the economic performance in the longer term, in order to reduce future maintenance and associated costs. The inconvenience which often arises during maintenance and other associated replacement costs, which may be out of all proportion to the costs of the part that has failed, also need to be examined. The different methods of procurement that are available may also make it easier and beneficial for the contractor to consider the effects of life cycle costing on a design.

At the construction stage

While the major input of life cycle costing is at the design stage, since its correct application here is likely to achieve the best in overall long-term economic savings, it should not be assumed that this is where the use of the technique ceases. At the construction phase there are three broad applications which should be considered.

The first of these concerns the contractor's method of construction, which, unless prescribed by the designer, is left to the contractor to determine. In some instances the contractor may be allowed to choose materials or components that comply with the specification but will nevertheless have an impact upon the life cycle costs of the project. The method of construction which the contractor chooses to employ can have a major influence upon the timing of cash flows and hence the time value of such payments. This is perhaps more pertinent to works of major civil engineering construction, where the methods available are more diverse. Buildability aspects which might enable the project to be constructed more efficiently, and hence more economically, may also have a knock-on effect in the longer term and, hence, have an influence upon the related costs-in-use. If involved sufficiently early in the project's life, construction managers are able to provide a professional input to the scrutiny of the design. They may be able to identify life cycle cost implications of the design in the context of manufacture and construction and in the way that the project will be assembled on site.

During the projects use and occupation

Life cycle costing has an important part to play in physical asset maintenance management. The costs attributable to maintenance do not remain uniform or static throughout a project's life, and therefore need to be reviewed at frequent intervals to assess their implications within the management of costs-in-use. Taxation rates and allowances will change and these can have an impact upon the maintenance policies being used. Grants may also become available for building repairs or to address specific issues such as energy usage or environmental considerations. The changes in the way the project is used and the hours of occupancy, for example, all need to be monitored to maintain an economic life cycle cost as the project evolves to meet new demands placed upon it.

When a project nears the end of its useful economic life careful judgement needs to be exercised before further expenditure is apportioned. The criterion for replacing a component is a combination of the rising running costs compared with the costs of its replacement and associated running costs. Additional non-economic benefits should also be considered and will need to be accounted for in some way in the analysis. For example, the advancement made in the improved efficiency of central heating boilers and their systems suggests that these, on economic terms alone, should be replaced every ten or twelve years irrespective of their working condition. A simple life cycle cost analysis is able to show that the improved efficiency of the burners and the better environmental controls will outweigh the replacement costs within this period of time.

In energy conservation

A major goal of the developed nations is towards a reduction of energy in all of its costly and harmful forms. This is true for the governments concerned, who have introduced taxation penalties, and for private industry who are seeking ways of reducing their own energy consumption and hence the associated costs. Life cycle costing is an appropriate technique to be used in the energy audit of premises. A reduction in energy usage has been encouraged due to the rising costs of foreign oil supplies, the finite availability of such fossil fuels and what has now become commonly known as the greenhouse effect. The energy audit requires a detailed study and investigation of the premises, recording of outputs and other data, tariff documentation and an appropriate monitoring system. The way that the premises are used plus typical or likely expectations of energy usage and sound professional judgements are important criteria for such an analysis. The recommendations may include, for example, providing additional insulation in walls and roofs, the replacement of obsolete equipment as well as suggesting values for temperature gauges, thermostats and other control equipment. An energy audit is not simply a one-off calculation, but one that needs to be repeated at frequent intervals in order to monitor the changes in the variables that affect the overall financial implications.

Table 5.1▶ Life cycle phases

Phase	Description	Cost implications
Specification	The formulation of the client's requirements at inception and briefing. Feasibility and viability of different proposals	Initial costs associated with land purchase, professional fees and construction
Design	Translating ideas into working drawings from outline proposals scheme and detail design	Cost planning including life cycle costing of alternative design solutions. Associated contract procurement documentation
Installation	The construction process	Interim payments and financial statements
Commissioning	Hand-over of the project to the client	Final accounts
Maintenance	The project in use	Recurring costs associated with repairs, running and replacement items
Modification	Alterations and modifications necessary to keep the project to a good standard	Costs associated with major refurbishment items
Replacement	Evaluation of the project for major changes or the site for redevelopment	Redevelopment costs

Source: Ashworth, 1999

5.4 Life cycle phases

The sequence of the seven life cycle phases is described appropriately in British Standard 3811:1974, and while this adopts engineering terminology that definition can also include the physical assets of buildings, as shown in Table 5.1.

5.5 Historical developments

The time value of money is not new. The pound in your pocket today is always worth more than it will be worth tomorrow. Life cycle costing is now the preferred term that is given to the evaluation of projects over their whole life span. The term originated in North America in the late 1950s. In the 1970s the US Government actively encouraged its use (Hoar and Norman, 1983). In the UK it was more incorrectly referred to as costs-in-use. It was not until much later that the UK began referring to life cycle costs (Flanagan and Norman, 1983). The strict definition of this term is the costs involved in purchasing and operating an asset. By definition this includes initial construction costs and even the costs associated with its eventual disposal. A RICS working party (RICS, 1986) prepared a report followed by a worked example (RICS, 1987). The engineering industry embraced the term 'terotechnology', which is a branch of technology and engineering concerned with the installation and maintenance of equipment. In other countries it has been known as total costs analysis, ultimate costs or whole-life costing. The use of discounted cash flow is a part of all of these techniques. This technique is able to convert costs in different time scales to a common time usually by expressing all the costs as present values or annual equivalents over the life of the building.

There is evidence that the varying techniques used in different parts of the world for evaluating the costs of buildings and other structures are seeking out more common ground. This is especially true of life cycle costing where similar studies have revealed similar problems in application. In most cases the overall approach is much the same. It consists, first, of examining different design proposals by assessing their respective costs, both initially and over their life cycle. Different countries, largely due to different traditions, carry out this process in different ways.

5.6 Deterioration and obsolescence

A distinction needs to be made between obsolescence and the deterioration of assets (Figure 5.1). Physical deterioration is largely a function of time and use. While this can be controlled to some extent by selecting the appropriate materials and components at the design stage and through correct maintenance while in use, deterioration is inevitable as an ageing process. Obsolescence is much more difficult to control since it is concerned with uncertain events such as the prediction of changes in design, fashion, technological development and

Figure 5.1▸ Obsolescence, deterioration and depreciation

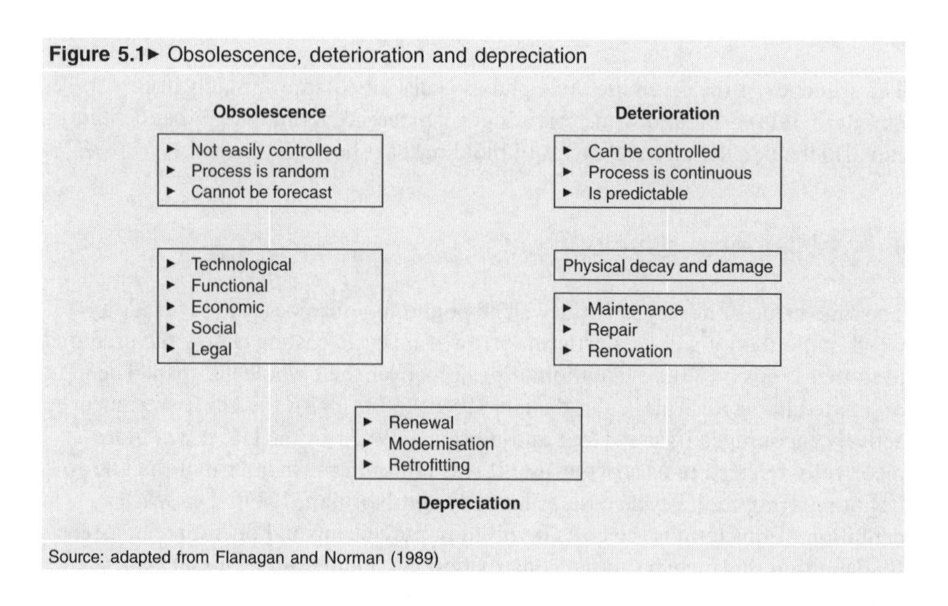

Source: adapted from Flanagan and Norman (1989)

innovation in the design and use of buildings. Deterioration eventually results in an absolute loss of use of a facility, whereas buildings that become obsolete accept that better facilities are available elsewhere. While deterioration in buildings can be remedied at a price, obsolescence is much less easy to resolve. Obsolescence can be defined as a decline in value that is not caused directly by use or the passage of time.

The word obsolescence has the following meaning: *That which is no longer practised or used, discarded, out of date, worn out, effaced through wearing down, atrophy or degeneration.* Such a definition relates to the decay of tangible and intangible things. All human products have an irresistible tendency to become old, but the speed of ageing is different for different objects and circumstances. Obsolescence is largely to do with changing requirements for which the object is no longer able to fulfil. For example, when existing standards of performance are replaced by new ones, functional obsolescence takes place.

Physical deterioration

Buildings wear out at different rates depending upon the type and quality of materials used and the standards and methods that were adopted for their construction. Ultimate physical deterioration is reached when a building is likely to collapse due to structural failure. However, in practice buildings rarely reach this stage since they are demolished prior to this occurring, normally for one of the reasons of obsolescence. Each of the various different components used within buildings has a different life span and each is capable of life extension or reduction depending upon the user's needs and the care exercised over its use. Where a building has been carefully designed and constructed, and properly maintained, its physical life can be extended.

The rates of physical deterioration of materials can be forecast within tolerable levels of accuracy using the lives of their respective building components. However, it must be recognised that extensive variation exists in the lives of even identical building components depending upon a wide range of different circumstances. These may include:

- An emphasis upon initial building costs without considering the consequences of costs in use.
- Inappropriate design and detailing of buildings and their components.
- Use of materials and components that have insufficient data on their longevity.
- Constructional practices on site that were poorly managed, supervised and inspected.
- Lack of understanding of the various mechanisms of deterioration.
- Insufficient attention given to the maintenance of the building stock.
- Inappropriate use by owners and occupiers.

Obsolescence

The life of a building may be considered in several different and distinct ways. Table 5.2 identifies the different sorts of obsolescence that designers and users need to consider. These are explained more fully in Ashworth (1999).

Table 5.2▸ Building life and obsolescence

Condition	Definition	Examples
Deterioration		
Physical	Deterioration beyond normal repair	Structural decay of building components
Obsolescence		
Technological	Advances in sciences and engineering results in outdated building	Office buildings unable to accommodate modern information and communications technology
Functional	Original designed use of the building is no longer required	Cotton mills converted in shopping units; chapels converted into warehouses
Economic	Cost objectives are able to be achieved in a better way	Site value is worth more than the value of the current activities
Social	Changes in the needs of society result in the lack of use for certain types of buildings	Multi-storey flats unsuitable for family accommodation in Britain
Legal	Legislation resulting in the prohibitive use of buildings unless major changes are introduced	Asbestos materials; fire regulations
Aesthetic	Style of architecture is no longer fashionable	Office building designs of the 1960s

Source: adapted from RICS, 1986

5.7 The main issues

During life cycle costing there are many difficulties that need to be resolved. Since it is a forecasting technique, and forecasts are invariably not precise, some expert judgement needs to be applied. The quality of the life cycle cost forecast is determined by the availability and reliability of the data and information that are used in the calculations and the skills that are employed by the practitioner when making judgements.

Building life

The life cycles of buildings are diverse during their inception, construction, use, renewal and demolition. There also lies a varied pattern of existence, where buildings are subject to periods of occupancy, vacancy, modification and extension. Using appropriate materials, components and technology, a building structure may be designed to last for about 100 years or more, depending upon the quality and standards expected from users. There are numerous examples of buildings that exceed this time span. However, the engineering services components in buildings have a much shorter life with an expectancy, at most, of about fifteen years and the life expectancies of finishes and fittings are now frequently less than ten years. By comparison, information technology hardware and software systems are becoming outdated even after a period of only three years.

The determination of building life expectancy is of fundamental importance in a life cycle cost calculation. However, in practice only limited consideration is given towards the assessment of building life expectancy at the time of its inception and design. By contrast, engineering systems are more carefully designed to meet expected and determined life cycle predictions. Materials and methods of manufacture, assembly and construction are selected on this basis to coincide with predicted life spans. Many industrial processes are reliant on life expectancy, often assuming that rapid changes in technology will make some processes obsolete and there is little point in attempting to design processes beyond their limited life spans.

The useful life of any building is governed by a number and coincidence of several different factors, including the sufficiency of the design, its constructional details and the methods used for construction on and off site. It is dependent upon the way that the building is used and the maintenance policies and practice undertaken during its life by its owners. It is also influenced by its location and the general ambience of adjacent properties.

There is a general shortage of informed data on the life expectancy of buildings and structures. There is also evidence that owners and users are unaware of either the total life span or the life expectancy up to renewal. They will have theoretical assumptions, but these are unlikely to mirror actual practice. There are examples of projects that have far exceeded their predicted life span and other buildings that have been demolished early because they were in the way of proposed developments. This latter scenario occurs where land prices are relatively high and commercial gain can be achieved.

The conundrum of predicting building life or life up to renewal remains unresolved. It has been recommended that for life cycle costing purposes the time scale should be the lesser of physical, functional and economic life. Sensitivity analysis can then be usefully applied to test the validity of life spans selected. Where the physical life span is the shortest, then this will be used as the basis. However, in practice this is rarely the case, with one of the different forms of obsolescence being of over-riding importance. Physical repair is possible in the majority of cases. It is more likely that one of the forms of obsolescence triggers the need for building renewal.

Component life

The life span of the individual materials and components have a contributory effect upon the life span of the building. However, data from practice suggest widely varying life expectancies, even for common building components. It is also not so much a question of how long a component will last, but of how long a component will be retained. The particular circumstances of each case will have a significant influence upon component longevity. These will include the original specification of the component, its appropriate installation within the building, interaction with adjacent materials, its use and abuse, frequency and standards of maintenance, local conditions and the acceptable level of actual performance required by the user. The management policies adopted by owners or occupiers are perhaps the most crucial factors in determining the span of a component's life, but there is a general lack and absence of such characteristics in retrieved maintenance data (Ashworth, 1996a, 1996b).

The design must recognise the difference between those parts of the building with a long, stable life and those parts where constant change, wide variation in aesthetic character and a short life are the principal characteristics. There seems to be little merit in including building components with long lives in situations where rapid change and modernisation is to be expected. All components have widely different life expectancies depending upon whether the physical, economic, functional, technological or social and legal obsolescence is the paramount factor influencing their life span.

An important and useful source of data for those involved in life cycle costing is their own accumulated research and expertise. In the absence of this, one or more of the published sources of information could be used. The data included (RICS/BRE, 1992) represents the findings from a questionnaire issued to a number of building surveyors. The information is typically represented in the format shown in Figure 5.2, which provides an indication of the sample size and the estimated component life in years using a variety of statistical measures.

Figure 5.2▸ Life expectancies of softwood windows and doors

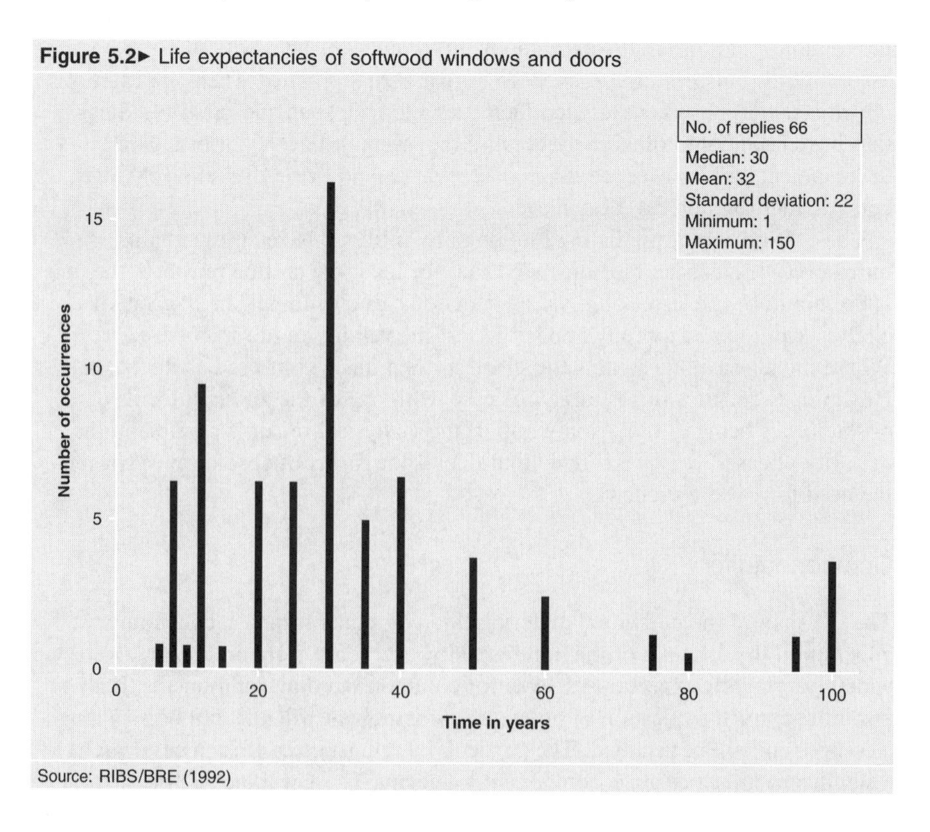

Source: RIBS/BRE (1992)

It can be observed from these data that the life expectancy of softwood windows and doors can vary between one and 150 years. Typically it represents a life expectancy of about thirty years. Furthermore, it would be foolish, for example, to prepare a life cycle cost based upon 150 years, even where guaranteed maintenance is promised, owing to the possibility of advancing obsolescence in buildings as identified above. Changes in use, the implications of fashion and the development of new technologies will also have some impact on life expectancies. The important message from investment analysts that 'past performance is no guarantee of future projections' can so easily be applied to building life cycles and the forecast of building component lives.

The survey does not provide an indication of the possible reasons for the expected different life expectancies. The replacement of the windows may be due to general decay, vandalism, fashion, the installation of double glazing, in order to reduce long-term maintenance, development of new technologies, etc. These and other data characteristics are not provided. If this information were included, then the range of values in a particular set of circumstances would be reduced. This would then allow its reuse in new situations to be made with greater confidence. On the basis of this and other information alone, it is not possible to select a precise life expectancy for a particular building component. Different techniques, such as sensitivity analysis or simulation, can be used to test the effects of best, worst, typical and any other scenario in terms of assessing the life expectancy.

The Housing and Property Manual (HAPM, 1995) Technical Note Number 6, *Life spans of Building Components*, provides yet further evidence of the variability of the lives of building component data. These are described more fully in HAPM (1992), *Component Life Manual*. The component lives indicated in this manual are based upon some general assumptions of good practice. In assessing the expected component lives a number of different factors are considered.

Sources of life-span data

A number of organisations, including research groups, professional bodies and manufacturers, provide information on building component life. However, an important point is that the prediction of component lives, for life cycle costing purposes, is not so much concerned with how long a component will physically last, but how long it will be retained. The scientific data are almost solely concerned with component longevity and not with obsolescence. While manufacturers and trade associations offer a valuable source of information, it needs to be remembered that the component life of a product may be described for ideal or perfect circumstances that rarely occur in practice.

Modes of component failure

The reasons identified for component failure are due to a combination and coincidence of several different factors. Researchers have recognised for a long time the vagaries associated with costs-in-use data. Even in similar circumstances, an identical component frequently has a different life span in practice.

Risk of component failure

The assessment of the life expectancy of individual components carries an element of risk. This is partially controlled through the long-term use of a component and its effective maintenance.

Practical experience

The use of the *Component Life Manual* (HAPM, 1995) is a good starting point and guide for those needing to assess the lives of building components, although the information is likely to be modified in practice by its users to suit the circumstances of particular buildings. It is also subject to revision on the basis of any past recorded data and the experiences of individual users.

Problems with component life data

There are a variety of different difficulties that have become associated with component life data. These include:

- Maintenance policies that are frequently driven by budget rather than by need.
- Data classification systems that are more akin to accounting than to forecasting.
- Component failure whose causes are rarely identified or reported accurately.
- Non-identical components used as replacement.
- Time lag delays in reporting problems.
- Hidden costs included within maintenance items.
- Potential distortion due to timing of maintenance operations.
- Knock-on effect of delayed maintenance work.

These issues are more fully explained in Ashworth (1999), which discusses the classification of component failure due to ageing and predictable wear and tear (replacement predictable), those that are due to accidental damage such as storm, vandalism and misuse (probability failure) and failure due to other components (associated items).

5.8 Costs and prices

Inflation

Throughout almost the whole of the twentieth century the UK has experienced erosion in the purchasing power of the pound. Much has been written on the causes and its possible cure. The effects of inflation and the problem that it causes to capital investment decisions need to be taken into account in a life cycle costing comparison. Even with relatively low levels of inflation, prices will be substantially affected over long periods of time, and the following are some of the characteristics of inflation:

- Inflation refers to the way that the price of goods and services tend to change over time.
- Inflation causes money to lose its purchasing power because the same amount buys less.
- The most commonly used measure of inflation in the UK is the Retail Prices Index (RPI). This is supposed to measure the costs of goods and services of a typical family's spending.
- The nominal rate of return on an asset or investment is the amount you get back; the real rate of return is the return after inflation has been taken into account.

- Cash deposits such as savings accounts, although secure, tend not keep pace with inflation.
- Interest rates are used to control inflation. By raising interest rates, governments can dampen consumer spending which results in reducing economic activity.
- Low inflation is supposed to be a good thing because it leads to price stability and may be internationally advantageous.
- The opposite threat of deflation is considered to be just as much a threat as inflation.
- Zero inflation is rarely desirable. The level of interest rates needed to achieve this would discourage economic activity.
- Europe measures inflation using an harmonised index. If this were adopted in the UK Britain's inflation record would look much better.

The principal problem facing the decision maker is whether to forecast future cash flows associated with an investment project in real terms or in money terms. 'Real terms' here means in terms of today's (the date of decision) price levels. 'Money terms' refers to the actual price levels which are forecast to obtain at the date of the future cash flow.

Two different approaches may therefore be used to deal with the problem of inflation. Firstly, inflation could be ignored on the assumption that it is impossible to forecast future inflation levels with any reasonable degree of accuracy. The argument is reinforced in that there is often only a small change in the relative values of the various items in a life cycle cost plan. Thus, a future increase in the values of the cost of building components is likely to be matched by a similar increase in terms of other goods and commodities. There is therefore some argument for working with today's costs and values. Also, since we are attempting to measure comparative values, real costs can perhaps be ignored.

However, changes in costs and prices and their interaction with each other are not uniform over time. Also, property values tend to move in booms or surges whereas changes in building costs are much more gradual. Costs do not necessarily increase in line with inflation. Reference to a range of different material or component costs over a period of time will show that these do not follow a uniform trend or pattern. Even similar materials, such as plumbing goods, can show wide differences over a ten-year cycle of comparisons. To ignore such differences will at least create minor discrepancies in the calculations.

The alternative approach in life cycle costing is to attempt to make some allowance for inflation within the calculations. This may be done, with some apprehension, using evidence of market expectations, published short- and long-term forecasts and intuitive judgements relating to the prevailing economic conditions. It is worth remembering that, in common with all forecasting, there will be a degree of error. The forecasting of inflation is a science and art in its own right. Mathematical models are constructed using a wide range of data to assist in their predictions. The models can only consider future events that may occur. In reality events occur that could not have been predicted even a few years earlier.

Discount rate

The selection of an appropriate discount rate to be used in the life cycle costs calculations will depend upon the financial status of each individual client. The discount rate to be selected will be influenced by the sources of capital that are available. The client may intend to use retained profits or to borrow from one of a number of commercial lenders.

It can be argued that the choice of a discount rate is one of the more crucial variables to be used in the analysis. The decision to build or to proceed with an investment may be influenced by the discount rate that is chosen. The selection of a suitable discount rate is generally inferred to mean the opportunity cost of capital. This is defined as meaning the real rate of return available on the best alternative use of the funds to be devoted to the proposed project. In practice the discount rate that is selected often represents the costs of borrowing, whether from the firm's own funds (loss of interest) or at a higher rate through borrowing. The discount rate that is proposed should then be adjusted by the expected rate of inflation or the time that the project remains live.

For simplicity, it is acceptable to arrive at a discount rate by deducting the expected rate of inflation from the cost of capital percentage. It is more correct to calculate the discount rate by using the following formula:

$$r = \left[\left(\frac{1+d}{1+i} \right) - 1 \right] \times 100$$

where r = net of inflation discount rate (real discount rate)
d = interest rate (cost of capital)
i = rate of inflation percentage

It is very important that, for each option being considered, the respective cash flows are calculated on exactly the same basis. If cash flows are to be estimated in nominal terms, i.e. include an estimate for inflation, they should be discounted at a nominal discount rate. This should then be applied to all the options under consideration. It is difficult to be definitive regarding which approach to adopt. Where cost estimates are assumed to inflate at the same rate, it is preferable to perform all calculations in current prices and to apply a real discount rate. However, where inflation is expected to operate differentially, the calculations should be done in nominal terms with explicit account being taken of the differential rates of inflation.

To select too low a discount rate will favour or bias decisions towards short-term, low-capital cost options. Selecting a discount rate that is too high will give an undue bias towards future cost savings at the expenses of higher initial outlays. The most accurate discount rate should reflect the particular circumstances of the project, the client and the prevailing market conditions. It is all too easy to tamper with the discount rate to make the calculation reflect the desired outcome. It is a matter of judgement, but one that is done within the context of best professional practice and ethics.

Taxation

Cash flows associated with taxation must be included in the calculation for practical reasons. Most project cash flows are affected by Corporation Tax. This may be due to capital expenditure attracting relief through capital allowances, profits from the project resulting in additional taxes, or losses attracting tax relief. Tax is not assessed by the Inland Revenue project by project but for the company as a whole. Cash flows must therefore be considered in this context and calculated on whether the project is carried out, delayed or abandoned. The matter is further complicated since the project may be spread over one or more tax years, where different taxation rates may need to be applied. Careful accounting may result in beneficial effects through tax avoidance measures.

Capital allowances are set against taxable profits in order to relieve the expenditure on fixed assets. There are several categories of asset into which statute has placed the various types of business fixed assets. Each of these has its own rules and basis for granting the allowance. In practice they are a combination of writing-down allowances and balancing charges.

Taxation relief varies, sometimes depending upon the type of building, and, in some cases, to encourage development of certain types of buildings. In the case of industrial buildings, for example, companies are able to deduct 4% of the cost of the building from the taxable profit in each year of its ownership and use. On disposal, the proceeds will cause a claw-back of excess allowances or additional allowance, if the difference between cost and the disposal proceeds has not already been fully relieved (see Chapter 11).

5.9 Sensitivity analysis

During a life cycle cost analysis a large number of different assumptions need to be made, such as building life expectancy and the longevity of the building components. While historic data may be used, the variability of such makes calculations difficult to perform. It is necessary, therefore, to test the judgements or assumptions in order to reduce as far as possible any distortions or misleading information that may have been introduced. One way of testing whether the results of the life cycle cost analysis remain stable under varying conditions is to repeat the calculations by changing the values that have been attributed to the individual variables. This is a technique known as *sensitivity analysis*. In practice, changing the values of the different variables used will result in changes to the overall outcome.

There are two different scenarios that can be employed. The first is to make changes to the life cycle cost model resulting from variations in the design, the materials used or the construction techniques to be employed. This will result in alternative life cycle costs being calculated. The preferred alternative design or construction solution may be obvious from these calculations; however, the results may be so similar that expert judgement needs to be applied in making the final design decision. The alternative approach is to provide life cycle cost models that test the stability of the model in varying circumstances. For example, the comparison of two alternative designs may reveal that design A is always preferable to design B in terms of the life cycle cost in normal circumstances, but the future might not necessarily be normal as we understand it today. The alternative approach is therefore to test the model at or even beyond the extremes of possibility to ensure that it remains stable under all conditions. This will rarely be the case in the real world and in different circumstances the alternative solution will be preferable.

It is therefore possible to test the effect on the life cycle cost of any variable used in the calculation. It needs to be remembered that life cycle costing is a technique to assist in the selection of the best or most economic of the alternatives that are available. Sensitivity analysis alone will not do this, but it will provide a range of different values that can be used to help an overall judgement to be formulated. Sensitivity analysis can also only be applied in known or expected circumstances. There are many examples where the future cannot yet be imagined or described.

5.10 Forecasting change

All forms of forecasting include elements of risk and uncertainty. They are estimates of future events and are, by definition, not precise. It has already been noted that some aspects are uncertain because they cannot be imagined or anticipated. All forms of forecasting suffer from such imponderables. Life cycle costing that relies on the future performance of buildings fits easily within this description.

Technology

It is difficult to forecast with any degree of accuracy the possible changes in technology, materials and construction methods that may occur in the near future. The construction industry, its process and its product are under a purposeful change and evolution. There is a constant striving to develop excellence in both design and manufacture and to introduce new technologies that have the desired characteristics of quality and reliability.

Fashion changes

These changes are much less gradual and more unpredictable than changes in technology and are influenced by society in general. Changes in the type and standards of provision, the use of space or the level of quality expectations can be observed from the past study of buildings (see section 3.5). Changes in the way that buildings might be used in the future have already been predicted. Some of these are hopelessly fanciful, while others reflect a changing attitude to work and leisure, changes in an individuals personal expectations, demographic trends and developments generally in society. Life cycle costing must attempt to anticipate future trends and solutions.

Cost and value changes

The erratic pattern of inflation throughout the past twenty-five years could not have been predicted even a few years earlier. The high inflation of the 1970s would not have been thought possible in the 1950s. An examination of building costs and tender prices show that these do not move in tandem. Economists have indicated that costs and prices cannot be assumed to rise indefinitely, and that there may be a future lapse or even a reversal of the traditional historic patterns.

Policy and decision-making changes

Perhaps one of the most important considerations is the future use and maintenance policy of the project by the owner. This factor is also the one that is generally an absent characteristic from the sparse historic data sources that are available. It has already been stated that much of the maintenance work is budget driven rather than needs oriented. Once the budget has been expended then no further amounts are available until the following year's budget allocations have been determined. The desire for proper maintenance of the physical asset is influenced by the costs and inconvenience involved. Different owners will set differing priorities and nothing can be assumed on the historical precedents of apportionments of other buildings. Studies emphasise that maintenance cycles and their associated costs must firstly be set properly within the maintenance objectives of the particular organisation concerned and the policies employed for planned and responsive maintenance.

5.11 Conclusions

The importance of attempting to account for future costs-in-use in an economic appraisal of any construction project has been established in theory. The question of whether it works in practice remains of crucial importance. However, the philosophy of whole cost appraisal is certainly preferable to the somewhat narrow and out-dated initial cost estimating approach. The widespread efforts so far expended in its research and development is a positive move, but further research is necessary to sharpen up the realities of the real problems encountered in practice. There is an eagerness to introduce new methods of evaluation without first being fully aware of all of the facts. Improving the education of those who are responsible for the design of capital works projects, and encouraging them to consider the future effects of their design and constructional details, remains an important consideration. Educating owners and users in how to obtain the best out of their building and engineering projects is another useful course of action to follow. The implementation of maintenance manuals or building owners' handbooks might also provide an improvement in the performance of buildings in use. The Private Finance Initiative (PFI) is also likely to encourage the use of life cycle costing.

Life cycle costing is at best a snapshot in time; in the light of present-day knowledge and practice and anticipated future applications. Some of the factors involved are of a crucial nature and can only be tested over a range of known values. Others are currently beyond our known expectations and may not even be considered as important factors today. Some of the assumptions may also be realised as untenable in practice.

There is some evidence from practice that life cycle costing is being more extensively used. This is particularly so in the case of major works projects, among clients who carry out an extensive amount of construction work and among those who are well informed of the consequences of design decisions on overall project costs. The following are some of the advantages of using life cycle costing and the added value that the technique can offer:

- The important emphasis that is placed on a whole life or total cost approach in the acquisition of capital cost projects or assets, rather than only a simple reliance on the initial capital costs alone.
- The ability to make a more effective choice between competing proposals within the stated objectives of a client. Life cycle costing takes into account the capital, repairs, running and replacement costs and expresses these in consistent and comparable terms.
- It provides different solutions for different design decisions and is able to set up hypotheses to test the confidence of the results achieved.
- It is an asset management tool that will allow the operating costs of premises to be evaluated at frequent intervals.
- It enables those areas of buildings to be identified as a result of changes in working practices, such as hours of operation, introduction of new plant or machinery, use of maintenance analysis, etc.

References and bibliography

Ashworth A. (1993) How life cycle costing could have improved existing costing. In J. W. Bull (ed.) *Life Cycle Costing for Construction*, Blackie Academic and Professional.

Ashworth A. (1996a) Estimating the life expectancies of building components in life cycle costing calculations. *The Surveyor*. Journal of the Institution of Surveyors, Malaysia. Fourth Quarter.

Ashworth A. (1996b) Data difficulties of building components for use in life cycle costing. *Journal of Structural Survey* 14 (3).

Ashworth, A. (1999) *Cost Studies of Buildings*. Addison Wesley Longman.

Flanagan R. and Norman G. (1983) *Life Cycle Costing for Construction*, Surveyors Publications.

Flanagan R., Norman G., Meadows J. and Robinson G. (1989) *Life Cycle Costing: Theory and Practice*, Blackwell Scientific Publications.

Hoar D. and Norman G. (1992) *Life cycle cost management*. In P. S. Brandon (ed.) *Building Cost Techniques: New Directions*, Blackwell Scientific Publications.

HAPM (1992) Housing and Property Manual: *Component Life Manual*, E. & F. N. Spon.

HAPM (1995) Housing and Property Manual, Technical Note No. 6: *Lifespans of Building Components*, E. & F. N. Spon.

RICS (1986) *A Guide to Life Cycle Costing for Construction*, Royal Institution of Chartered Surveyors. Surveyors Publications.

RICS (1987) *Life Cycle Costing: A Worked Example*, Royal Institution of Chartered Surveyors.

RICS/BRE (1992) *Life Expectancies of Building Components. Preliminary Results from a Survey of Building Surveyor's Views*, RICS Research Paper No. 11. Royal Institution of Chartered Surveyors and the Building Research Establishment.

Chapter 6

Value management

6.1 Introduction

Value management emerged from the demands of the USA manufacturing industry in the Second World War. The concept of 'value analysis' was developed by Lawrence Miles, an electrical engineer with the General Electric Company, who adopted a functional approach to the purchasing requirements of his company. This involved the analysis of a component part of a product in terms of the function it performed and the search for an alternative solution to provide that function (rather than the product) at lower cost. Throughout the 1940s and 1950s the use of the concept further developed and expanded within the USA, becoming a procedure which could be used during the design or engineering stages. In 1954, the term 'value engineering' was introduced by the US military. During the 1960s value engineering spread to the UK manufacturing industry and in the USA it was introduced to the construction industry. The concept of 'value management' was first used within the UK construction industry in the 1980s. While value techniques are now used by manufacturers on a global scale, their application within the construction sector is mainly found in the USA, the UK and Australia. The position of value management is now established by legislation in both the USA and New South Wales, Australia. In the UK, several bodies have issued documents providing recommendations and guidelines; however, the Government has stopped short of any mandatory requirement.

There are several value management societies in existence throughout the world which aim to serve value managers by promoting, developing and controlling standards of practice and qualifications. In the UK, the Institute of Value Management introduced a programme of formal training in September 1998 that allows practising and prospective value managers to obtain recognised education and training leading to a UK qualification – Certified Value Analyst or Certified Value Manager.

Practitioners in value management, and clients that have adopted its use within their organisations, testify to its success. The list of clients using value management in the UK contains many major organisations, for example

Railtrack, London Underground, BAA, BNFL, and the promotion of the practice at a high level has continued for the past decade. Value management has been recognised as a factor that is critical to the success of projects, by providing the basis for improving value for money in construction. It therefore provides practitioners with an opportunity of contributing further to the added value service they provide.

6.2 Terminology

To those being introduced to the subject for the first time, it is important to note that there is quite a wide variety of value management terminology used which may or may not mean anything in particular and thus can be confusing. For example, frequently the terms value management, value engineering and value analysis are used synonymously to mean the entire concept. Alternatively, value engineering is often considered as defined below and value management is often intended to include value planning, also defined below. To fully outline the varied use of terminology would require more attention than is considered worth while; the semantics are unimportant provided the meaning of each term in context is understood, and readers carrying out further research into this subject are advised to bear this in mind. The following definitions are suggested for the sake of clarity and the understanding of any reference in this text.

- **Value management**: This is the term used to describe the entire philosophy and range of techniques. Therefore, value planning, value engineering and value analysis form a subset of value management (see Figure 6.1).
- **Value planning:** This, as the term may suggest, is carried out in the early part of a project prior to the decision to build or at briefing or outline design stage. Value techniques are used to reach a group decision in terms of the criteria for a proposed design or the criteria for a business strategy. It is possibly a common misconception that the use of value techniques is intended for problem solving (as in 'value engineering') – typically for the remedying of a budget over-run – following the production of a scheme design. This is not the case.
- **Value engineering**: This is the term used to describe value techniques that are adopted during the detailed design stages and construction stages when completed designs or elements of the design will be available for study.
- **Value analysis**: This is the term used to describe value techniques that are carried out following the completion of a building.

Figure 6.1► The value management 'family'

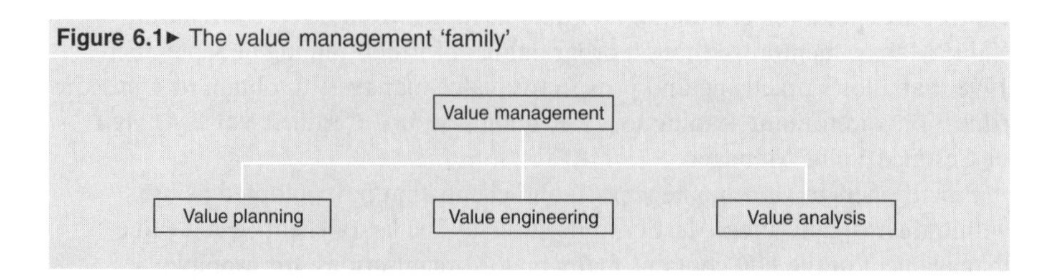

6.3 The value management process

Many texts on value management will focus upon definitions of 'value' or 'cost' or 'worth' and describe various systems of application, all of which are necessary, but which may, to some, have the appearance of strict disciplines or sciences to which they must rigidly adhere. Value management is a simple concept that primarily involves a workshop, or a series of workshops, in which professionals and client representatives search for better value in terms of a client's objectives. The process and tools and techniques used during the workshop can determine whether or not the exercise is one that can be classed as value management. Norton and McElligott (1995) propose a test for 'authentic value management': 'For a service or activity to be characterised as authentic value management, there are four criteria which must be satisfied.' The exercise must:

- '. . . follow an approved value management job plan . . .'
- '. . . involve a multi-disciplinary team . . .'
- '. . . be facilitated by a qualified value manager . . .'
- '. . . not pursue any design changes which detract from the project's required or basic functions . . . generally assured through the formal application of function analysis techniques . . .'.

These criteria are considered essential to 'authentic' value management and are now discussed.

The job plan

The process of value management involves a structured workshop, or series of structured workshops at which group decisions are made. An important aspect of the value management workshop is its structure, which usually follows a five-phase process known as the 'Job Plan'. This structured approach was developed by Lawrence Miles and, while academics and practitioners have refined and contextualised a variety of differing methods, it is normally adhered to in all approaches to value management in some form. While this suggests a rigid approach, it should be regarded as an outline; situations and projects will differ and make their own demands on workshop approach and activity. For example, a workshop may be limited to the information, speculation and evaluation phases (described below) where time is restricted or resources are not available with provision for development and presentation in a follow-up meeting. Likewise, the structure shows a logical and sequential path which, in practice, may vary and necessitate iterative action. The stages of the Job Plan are discussed below:

The information phase

The workshop commences with an *information phase* in which details of the problem or project are presented to participants. If the study relates to a proposed building, this may include a presentation by the client representatives, the architect, structural engineer, quantity surveyor and other design consultants which will provide details of project background, objectives (distinguishing between a clients needs and wants) and constraints (e.g. site, budget, time). While the primary aim of the information phase is to provide all participants with an understanding of the project, it also serves as a team-building exercise which is useful in preparing a fertile platform for the remainder of the workshop activities. A feature of this phase of the workshop is functional analysis, which is frequently carried out via the production of a functional analysis diagram (see section 6.5). This leads to a better understanding of the project and allows the identification of those items which are considered 'poor value' in terms of cost v. worth (see below) which will form the focus of further study.

The creative phase

Once the project participants have a thorough understanding of the project, including an appreciation of its functional requirements, the workshop team is invited, in the *creative phase*, to generate alternative solutions and ideas. This phase of the workshop is usually performed with the aid of brainstorming and other creative thinking techniques to encourage the contribution of suggestions to improve value. With the application of the key brainstorming rules – that as many ideas are generated as possible and that no opinion regarding any of the ideas is attempted or given until the creative phase is complete – it is common to produce a large number of suggestions for future evaluation.

The evaluation phase

There are several methods that may be used during the *evaluation phase* to assess the merits of the ideas generated during the creative phase. A problem to be faced at this stage is how best to obtain the agreement of workshop participants on the selection of ideas for further development. The choice of approach will be situation dependent (e.g. influenced by time available, timing of the workshop, composition of workshop team, complexity of project) and may rely on a democratic procedure or the ability of the facilitator to obtain open consensus. One example of a specific evaluation technique is 'championing', which is reliant upon team members volunteering to 'champion' particular ideas, i.e. accepting responsibility for their development and subsequent implementation or reasoned rejection. Thus, ideas without 'champions' are rejected. Whichever method is used, a list of best ideas is then carried forward for further development.

The development phase

While an evaluation of ideas generated in the creative phase has occurred by this stage of the workshop, this has probably been based upon no more than an outline perception. The *development phase* allows for further work that is necessary to determine whether or not an idea should become a firm proposal or, if the workshop is at briefing stage, to consider the incorporation of ideas within a revised brief. This workshop phase is time consuming and will probably involve significant technical input. It is nevertheless essential to carry out this developmental work prior to presenting formal proposals. Because of the time-consuming nature of the development phase, it is often beneficial to complete the associated work beyond the confines of the workshop and present the results at a subsequent meeting.

The presentation phase

The objective of the *presentation phase* is to present the team's proposals to the client representatives. The presentation of proposals, which will probably include adjustments to original design proposals, is something that may be very sensitive to consultants and possibly the client.

A written report should be prepared by the facilitator which will incorporate details of the proposals. This will also include a plan of action for their implementation.

6.4 Multi-discipline participation

Value management is a *multi-discipline* exercise in that it ideally requires the participation of consultants from all relevant design disciplines and client representatives who share a common interest (and thus are *stakeholders*) in the success of the project. To be effective, and as it is regarded as critical to the success of value management, the team should have an appropriate mix of experience, knowledge and skills and, dependent upon workshop objectives, a range of stakeholder perspectives. In a recent value management study for a large university library project, the workshop, which was managed by two facilitators, included the following team members:

- Consultants (all with existing involvement other than the contractor): two architects (with differing design responsibilities); two structural engineers; one services engineer; one contractor (acting as an independent adviser).
- Client representatives: library staff; information technology specialists; teaching and learning managers; estate managers.

With a core workshop membership of fourteen, this may be regarded as too many; however, the two-day exercise was well managed and proved very effective with positive feedback from all participants.

An important consideration when selecting a value management team is whether to use existing design team members or an independent workshop team. There are several advantages to the use of the existing design team. Where existing project members are used, there is less likelihood of difficulties with the implementation of 'outsiders' ideas, costs are curtailed, there is a saving in time due to the existing knowledge state of the project and it could prove to be a useful team-building exercise. However, a potential major disadvantage could be the dominance of original design concepts, if these have been established, which may be strongly defended by designers. The prospect of this occurring is supported by research into design practice and it is something that facilitators need to be aware of. With the exception of value management practice in the US public sector, the participation of independent design consultants would appear to be uncommon.

Facilitator

The ability of the *facilitator* is central to the success of the value management process. The role of the facilitator may include advising upon the selection of the value management team, co-ordinating pre-workshop activities (e.g. issue of relevant information to selected value management participants), deciding upon the timing and duration of workshops, managing the workshop process and preparing reports. The management of the workshop can be a difficult task requiring a variety of skills, including: the ability to adhere to an agenda; identifying the strengths and weaknesses of team members and promoting their positive interaction; motivating and directing activity; overseeing functional analysis; promoting an atmosphere conducive to creativity while at the same time maintaining a disciplined structure.

Other critical success factors

In addition to the characteristics considered essential to the practice of 'authentic' value management described above, other factors are considered critical to its success. These include: the support of *senior management*; an appropriate workshop *location*; the correct *timing* of the value management study; the presence of *decision takers*; the completion and monitoring of an *implementation* plan to ensure that post-value management study action is taken; and adequate *pre-workshop* preparation.

6.5 Functional analysis

There is mixed opinion as to the importance of functional analysis within construction-related value management. The benefits of functional analysis are well recognised and considered by some to be central to the value management process. This opinion is not universal, however, and there are many US practitioners who consider 'its structured use of little value in the performance of a construction-orientated value management exercise' (Kelly and Male, 1993).

The underlying principle of functional analysis is both simple and effective. When applying the technique to a building component or element, it invites the question 'What does it do?' as opposed to 'What is it?'. Thus, when searching for alternatives, we look for something that will perform the required function rather than attempt to find a substitute for the previous solution.

To demonstrate some of the principles and benefits of functional analysis, reference to the matrix in Figure 6.2 is used to show a hypothetical analysis of the costs of a softwood window installation in terms of functional requirements. It should be noted that the values and function allocation indicated are entirely notional and are there purely to serve the explanation. The logic behind the functional costing is related to *cost* v. *worth* and makes the convenient assumption that all costs can be allocated to some function. Worth is defined as the least cost necessary to provide the function. Thus if we focus upon the functional costs of the 'casement', the following can be observed:

- The minimum cost of a casement to serve the basic functions of 'control ventilation', 'exclude moisture' 'retain heat' and 'transmit light' is £60 (assumed to be present in *all* windows). This amount has been allocated in equal amounts to each of the functions.

Figure 6.2▸ Functional matrix: 'softwood window'

Component \ Function	Primary					Secondary						
	Permit ventilation	Control ventilation	Exclude moisture	Retain heat	Transmit light	Improve security	Reduce sound	Reduce glare	Extend life	Assist cleaning	Enhance appearance	**Component cost (£)**
Lintol	15				15							30
Opening	10				10							20
Frame		5	5	5	5				10		5	35
Casement		**15**	**15**	**15**	**15**				**30**		**35**	**125**
Ironmongery		10				5				20	5	40
Glass		5	5	5		5	10	5			10	45
Paint									5		10	15
Function cost	25	35	25	25	45	10	10	5	45	20	65	**310**

- An additional cost of £30 is attributed to the increased specification of softwood – the function of which is to 'extend life' (and reduce the maintenance) of the window.
- An additional cost of £35 is attributed to the window style, say, of Georgian appearance in small panes incorporating moulded sections – the function of which is to 'enhance appearance'.
- Function is not always as it first may seem. The function of 'permit ventilation' is achieved by forming an opening; with regard to ventilation, the purpose of the casement is to 'control ventilation'.
- The functions can be divided into *primary*, i.e. those which are essential, and *secondary*, i.e. those which are not essential (but possibly unavoidable), and are often provided in response to the design solution (e.g. reduce glare). The choice of what is basic or secondary may be subjective and dependent upon individual perception (hence the need for *stakeholder* participation). The separation of basic and secondary functions assists with the understanding of a project and may identify areas to target for value improvement.
- In terms of cost v. worth, if the basic functions only are required as shown (as stated above this is a subjective view), the window is only *worth* £165 (the sum of the primary function totals). Thus, if we have expended £310 and don't require to 'reduce sound', 'reduce glare', extend life', 'assist cleaning' and 'enhance appearance' (e.g. we live in a bungalow, in a quiet rural hamlet, surrounded by trees and have simple aesthetic tastes) we have not achieved good value.

This demonstrates how functional analysis can be used to assist with the identification of *unnecessary costs* or to highlight an imbalance of expenditure that could lead to substitution with a design alternative. In practice, the benefits of functional analysis on a component such as a softwood window may not prove worth while due to its probably small relative value and the possible limitation on the provision of alternatives; however, the benefits of the technique can be seen.

6.6 Functional analysis diagramming techniques

Functional analysis diagramming techniques can be used across the varying levels of a project development – for example: at strategic stage to possibly identify the building need; at built solution level to assist in the briefing process; at elemental level to identify unnecessary costs. In practice however, the use of the technique at technical level is doubtful. Examples of these are shown in Figure 6.3.

Figure 6.3▸ The use of How/Why logic

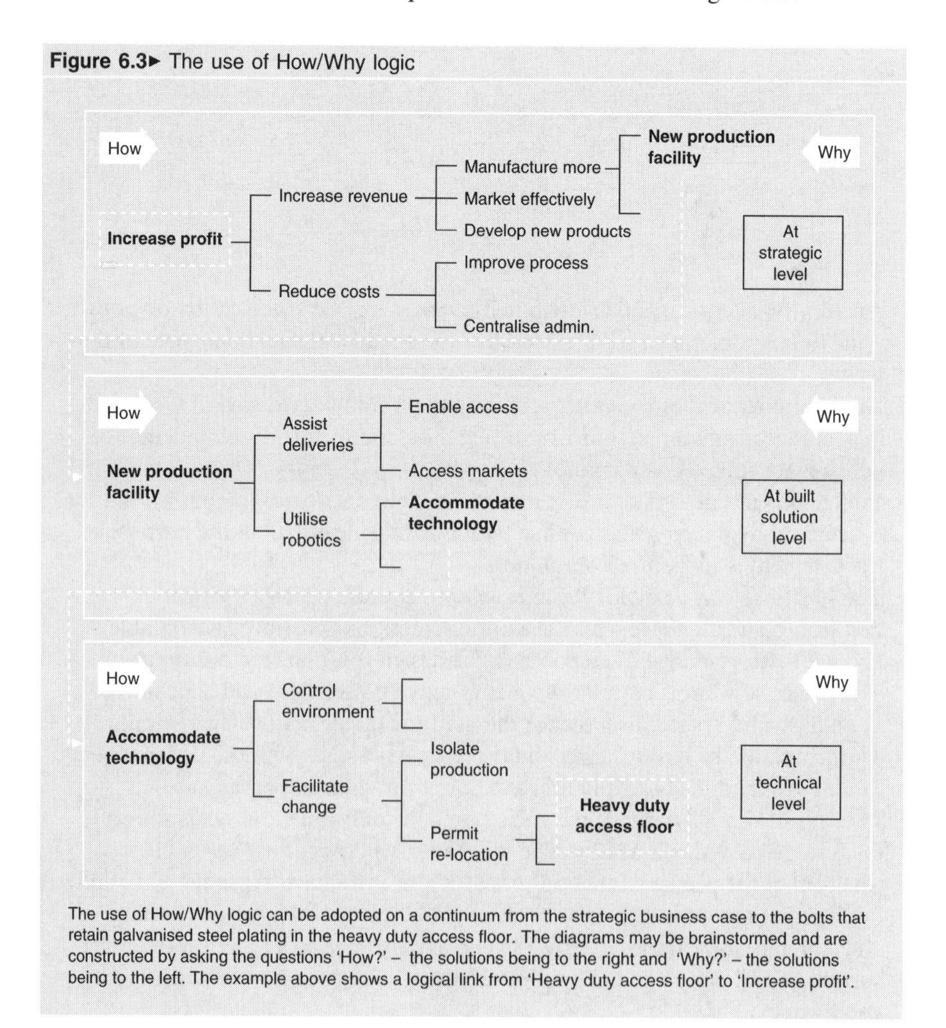

The use of How/Why logic can be adopted on a continuum from the strategic business case to the bolts that retain galvanised steel plating in the heavy duty access floor. The diagrams may be brainstormed and are constructed by asking the questions 'How?' – the solutions being to the right and 'Why?' – the solutions being to the left. The example above shows a logical link from 'Heavy duty access floor' to 'Increase profit'.

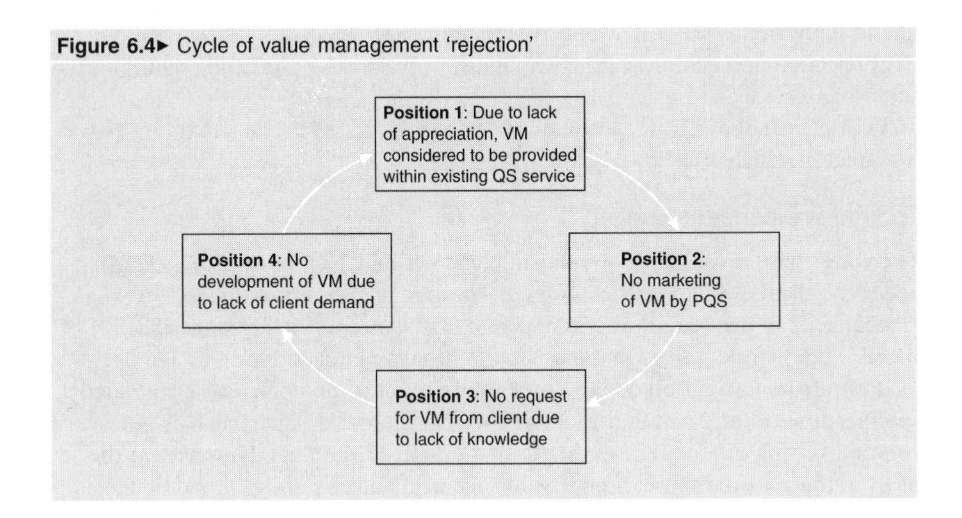

Figure 6.4► Cycle of value management 'rejection'

6.7 The case for value management

General claims of the success of value management in financial terms suggest that for a value management fee of 1%, a 10–15% cost saving can be achieved. It is hard to imagine that practitioners and clients would not use value management in the belief that this level of result could be attained. However, general and vague statements such as the one above are difficult to prove, perhaps particularly to quantity surveying practitioners who frequently achieve the same order of cost savings for clients in their traditional role.

Despite the acclaim given to value management, as far as the quantity surveyor is concerned, it is used on less than 10% (probably much less) of the projects on which they are involved. This may be due to several reasons including lack of client demand, client's reluctance to pay an additional fee, or a belief that the quantity surveyor provides the service already. A possible cycle of value management 'rejection' (Hogg, 1999) is shown in Figure 6.4.

- **Position 1**: Due to a lack of understanding, the quantity surveying practitioner considers that the benefits of value management are inherent within the existing service.
- **Position 2**: As a result of this belief, the quantity surveying practitioner doesn't actively promote value management as a distinct service.
- **Position 3**: Since value management isn't promoted by the quantity surveyor, the client isn't aware or doesn't appreciate the benefits and therefore doesn't demand the service.
- **Position 4**: Since the client doesn't demand value management a level of client satisfaction in the existing service is assumed; the quantity surveyor doesn't develop value management skills, resulting in a return to position 1.

Because of this cycle, it is possible that the benefits of value management will be kept from a significant element of the client body. It is therefore expedient to state further the case for value management, not by making claims as to the financial rewards it may offer, but by explaining why poor value is reduced by its application and demonstrating why it differs from the traditional quantity surveying service.

As discussed above, value techniques can be applied in several situations, two examples of which are shown below.

To improve concept briefing

There are many causes of poor value in construction which most professionals, and some clients, will have encountered. The main causes occur due to inadequacies in the design process, including the concept briefing stage. The list of possible poor value factors that may have a negative impact during design could incorporate limitations on design time forcing error or incompleteness, outdated specifications, poor communications between client and designer such as poor concept briefing or a poor co-ordination of design consultants. However, at the heart of much of the cause of poor value occurring during design could be the nature of the design process itself. Owing to the complex nature of construction design, architects adopt a design strategy that results in the early production of sketch proposals. Research suggests that the early designs, often based upon relatively little information, are dominant in determining the nature of the final design. Since it is generally considered that 80% of costs are committed at concept design, the opportunity to add value at this stage is significant.

Value planning is carried out in the early part of a project prior to the decision to build or at briefing or outline design stage. This provides the opportunity for a multi-discipline team, including client stakeholders, to examine project objectives and, with the use of value techniques, to reach a group decision in terms of the criteria for a proposed design. One of these techniques may be the production of a functional analysis diagram and the example above indicates the use of this technique at strategic level. This diagram, which can be weighted to indicate design priorities, will assist in directing members of the design team towards the project objectives and may be used to monitor design output. Value management can therefore be used to mitigate the inadequacies of traditional concept briefing.

To attain savings

A frequently heard comment made by quantity surveying practitioners is that they practise value management as part of their traditional role in that part of their function concerns the generation of cost reductions. This is a misunderstanding; the cost reduction exercise that is usually carried out by the quantity surveyor is done with the objective of reducing costs without regard to function. Value management focuses upon eliminating *unnecessary costs* (i.e. costs which provide no function). This is demonstrated by the following examples:

At tender stage, a proposed library building is over budget by, say, £250,000:

Traditional cost reduction exercise by a quantity surveyor

The quantity surveyor considers that the internal finishes element appears to offer scope for savings and proposes, for consideration, a schedule of cost reductions. This includes a reduction in the specification of floor tiles (the substitute having a reduced life span, being more difficult to maintain and having a less attractive appearance) and the omission of access flooring from two floors. Clearly, the required budget can be achieved by this action, but with a loss of function that the client would have preferred to retain, even if not needed.

Value management approach

A facilitator co-ordinates a value engineering workshop with the objective of achieving savings of £250,000. The workshop team, using value techniques, ascertains that the function of 'accommodate staff' has been provided by the inclusion of personal office space for all members of staff which has been based upon previous library practice. The workshop team find that the cost of this function (£750,000) can be halved, without loss of function, since, with the increased use of IT and 'hot desking', private staff accommodation can be provided on a staff ratio of 33%. It may be observed that if value management had been used early in the design process, this element of poor value would probably not have occurred. Since it is likely that the retrospective improvement to design is less efficient than 'getting it right first time', the beneficial use of value management early in a project's life is emphasised.

The difference between cutting costs and cutting unnecessary costs is fundamental to the value management approach.

References and bibliography

Green S. D. (1992) *A SMART Methodology for Value Management*, The Chartered Institute of Building, Occasional Paper No. 53.

HM Treasury; Procurement Guidance No. 2: *Value for Money in Construction Procurement*, HMSO.

HM Treasury (1996) *Central Unit on Procurement Guidance Note No. 54: Value Management*, HMSO.

Hogg K. I. (1999) Value management; a failing opportunity? *Conference Proceedings; COBRA 1999*, Royal Institution of Chartered Surveyors.

Kelly J. R. and Male S. P. (1988) *A Study of Value Management and Quantity Surveying Practice*, Occasional Paper, Surveyors Publications, London.

Kelly J. R. and Male S. P. (1990) A critique of value management in construction. *CIB W55*, Sydney.

Kelly J. R. and Male S. P. (1991) *The Practice of Value management: Enhancing Value or Cutting Cost*, Royal Institution of Chartered Surveyors.

Kelly J. and Male S. (1993) *Value Management in Design and Construction: The Economic Management of Projects*, E. & F. N. Spon.

Latham, Sir M (1994) *Constructing the Team*, London, HMSO.

Male S., Kelly J., Fernie S., Grönqvist M. and Bowles G. (1998) *Value Management – The Value Management Benchmark: A Good Practice Framework for Clients and Practitioners*, Thomas Telford.

Norton B. R. and McElligott W. C. (1995) *Value Management in Construction: A Practical Guide*, Macmillan.

RICS (1995) *Improving Value for Money in Construction: Guidance for Chartered Surveyors and Clients*, University of Reading for Royal Institution of Chartered Surveyors.

Smyth G. (1999) Dare surveyors ignore value management? *The Chartered Surveyor Monthly*, February, pp. 46–47.

Chapter 7

Partnering

7.1 Introduction

Many of the problems that exist in the construction industry are attributed to the barriers that exist between clients and contractors. In essence, partnering is about breaking down these barriers by establishing a working environment that is based upon the mutual objectives of team work, trust and sharing in risks and rewards.

Within the UK construction industry, partnering activity is a relatively recent phenomenon being given significant impetus by the Latham Report (Latham, 1994) and many subsequent publications and positive action. The University of Reading report for the Royal Institution of Chartered Surveyors, *Improving Value for Money in Construction: Guidance for Chartered Surveyors and Clients* (RICS, 1995) was prepared in recognition of Sir Michael Latham's report *Constructing the Team*. This highlights the benefits of partnering as one factor that is critical to the success of construction projects in providing the basis for improving value for money in construction. The report stated that:

> 'An attitude of co-operation amongst the project team members and, not least, with the client must be created. Partnering can help enormously without running the risk of being uncompetitive or introducing complacency.'

Partnering also featured prominently in the report by Sir John Egan's Construction Task Force, *Rethinking Construction* (Egan, 1988). This was regarded as central to improving the performance of the construction industry. The prominent recognition given to partnering in the last five years of the twentieth century is a clear indication of the strong belief that the partnering approach can make a major contribution to improving or adding value within the construction industry.

Throughout this chapter, reference to partnering participants is generally restricted to client and contractor. The relationships and implications of partnering can apply and are encouraged to apply throughout the supply chain and readers should bear this in mind accordingly.

7.2 What is partnering?

Partnering is reliant upon the principle that co-operation is a more efficient method of working than confrontation. This is frequently the approach that results from traditional contracting in which each party is driven towards looking after their own independent objectives. These principles are simply and well described by Bennett (1999), who stated that:

> 'Firms are better off when they work to make the cake bigger than when they fight to get a bigger share of the existing cake. This is true as long as they make sure that everyone gets a fair share of the cake.'

From this, the intentions of partnering are clear, as is the indication that fairness and trust play a major and important role. Traditionally, a great deal of energy is expended on retaining as much of the cake as possible and little to prompt efficiency and improvement. Partnering aims to remedy this.

There is no universally accepted version of partnering and the range of definitions clearly demonstrates this. The version below (which, it is important to add, has been subsequently revised by its authors) provides a relatively simple view, which should be easily understandable by all:

> 'Partnering is a management approach used by two or more organisations to achieve specific business objectives by maximising the effectiveness of each participant's resources. The approach is based on mutual objectives, an agreed method of problem resolution and an active search for continuous measurable improvements.' (Bennett and Jayes, 1995)

It can be seen from this definition that there are three main components to a partnering arrangement:

- **Mutual objectives**: These should be accepted by all partners at commencement of the project and reviewed if necessary during the project duration. It is common to incorporate these within a *charter* that may be publicly displayed to remind parties continually of their obligations.
- **An agreed method of problem resolution**: A better way of resolving disputes should be established based upon fairness and the willingness to find win–win solutions. These should be outside the usual contract methods.
- **An active search for continuous and measurable improvements**: This is particularly relevant to strategic partnering arrangements where continuous improvement is considered as a major benefit. Benchmarking is a method of monitoring this, using both internal and external sources for comparison (see Chapter 8 on Benchmarking). This principle also fits easily within the aim of adding value in the construction of projects.

7.3 Categories of partnering

The nature of partnering is such that it may take several different forms dependent upon individual situations and the objectives of the various parties involved. However, it is possible to broadly classify a partnering arrangement as either *project partnering* or *strategic partnering*. While the differences between these two classifications, which relate to scale and level of relationship, are significant, the essence of the partnering concept is the same in both.

Project partnering

As the name would suggest, project partnering relates to a specific project for which mutual objectives are established and the principles involved are restricted to the specified project only. The great majority of partnering opportunity is of this type since:

- *It can be relatively easily applied in situations where legislation relating to free trade is strictly imposed.* In the UK, a *Post-Award Project Specific Partnering* methodology has been proposed and has been published by the European Construction Institute (ECI, 1997). This is seen as particularly appropriate to the public sector since:

  > 'It allows an openly competitive process of selecting contractors to be adopted. Thus the European Union's public procurement requirements are respected. It is also fully compatible with the requirements of compulsory competitive tendering (CCT) as imposed under the various local government Acts.' (ECI, 1997)

- *Clients seeking to build on an occasional basis may use it.* The vast majority of clients who, due to the size and timing of their development programme, are not in a position to enter into a long-term partnering relationship.

The predicted rewards of project partnering are much less than where longer term strategic partnering arrangements exist. However, evidence from the USA, where approximately 90% of partnering is of the single project type (as a result of difficulties relating to the competition law), shows that benefits may be obtained (Bennett and Jayes, 1995).

Strategic partnering

Strategic partnering takes the concept of partnering beyond that outlined for project partnering to incorporate the consideration of longer term issues. This is an important aspect that should be considered before assuming the existence of strategic partnering. In many cases in practice, what may be identified as strategic partnering is effectively a project-based approach used on each of a series of projects. This is done without the added requirement of long-term strategic considerations (Bennett and Jayes, 1998).

The additional benefits of strategic partnering are a consequence of the opportunity that a long-term relationship may bring and could include:

- *Establishing common facilities and systems.* The use of shared office accommodation and communication access and storage systems promotes openness, efficiency and innovation.
- *Learning through repeated projects.* Construction processes can be developed to reduce defects and improve efficiency leading to additional cost savings, reductions in the time required for construction and an improvement in quality.
- *The development of an understanding and empathy for the partners longer term business objectives.*

7.4 The partnering process

There are a variety of versions of the partnering process, some of which appear quite complex and are continuing to develop to meet particular objectives. It is not considered important or possible in this book to discuss these in depth and readers are referred to the growing number of texts that cover this aspect in more detail. Once the decision to partner has been made, and partnering may not be an appropriate method to adopt, the procedure of partnering principally involves:

- **A selection procedure**: The selection of a good client or contractor partner who is trustworthy and committed to the arrangement is fundamental to the process. Partnering can be used with the traditional methods of procurement in the initial selection stages.
- **An initial partnering workshop**: This will be attended by the key stakeholders and usually results in the production and agreement of a partnering charter which will be signed by all participants (see Figures 7.1 and 7.2).
- **Project review**: During the project implementation stage, performance will be regularly reviewed. This will incorporate all relevant project matters including, for example, quality, finance, programme, problem resolution and safety. The review period and format, which may include further workshops, will be dependent upon the needs of the partnering objectives.

7.5 The attributes and concerns of partnering

Attributes

The adoption of partnering, at a strategic level or for a specific project, is considered to bring major improvements to the construction process resulting in significant benefits to each partner. These include:

- **Reduction in disputes**: Disputes within the construction industry abound. Adversarial relationships may co-exist between clients and contractors, contractors and subcontractors, subcontractors and suppliers, consultants and clients, consultants and contractors or subcontractors. In fact, it is possible that difficulties may arise between any combination of a project's participants. In settling disputes, the traditional solution usually leads to success for one party and failure for the other. As outlined above, this situation may impact throughout the supply chain and possibly, within various relationships, simultaneously on a specific project. This negative outlook provides a strong argument in favour of partnering which has been shown to reduce the number of disputes.
- **Reduction in time and expense in the settlement of dispute**: The occurrence of disputes is unfortunately an accepted reality of the construction industry. Partnering does not eliminate them and they are provided for within the partnering framework. However, the framework provides for faster and more efficient methods of resolution.
- **Reduction in costs**: The key attributes of partnering include the beneficial effects of repetition, improvements in communications, innovation in both

Figure 7.1▶ Partnering charter agreement: Railtrack/Battersea Depot

RAILTRACK
the heart of the railway

BATTERSEA DEPOT

CHARTER

Our aim is to successfully complete the project enabling Gatwick Express to continue with their existing rolling stock maintenance and the introduction of the new Class 460 stock.

Objectives:

- To operate in a safe manner at all times
- To develop and maintain effective communication between all parties
- To implement a right first time culture
- To plan and execute the work in order to minimise disruption to the depot operations
- To minimise the need for out berthing of new trains
- To develop and maintain a good working relationship built on trust
- To enable the works to be carried out to the agreed programme
- To regularly liaise to ensure agreed delivery
- To maintain a safe working environment for all parties at all times
- To ensure prompt acceptance of works enabling prompt payment
- For all parties to make a profit from excellent team performance

GATWICK EXPRESS

Birse

23 March 1999

Source: Birse Rail/Railtrack; reproduced with permission

design and construction and the search for continued improvement. This approach can result in significant savings in construction cost. Research has shown that the type of partnering utilised can have a major effect on the scale of cost savings, as demonstrated by the research findings of the Reading Construction Forum:

> 'Typically, with project partnering, cost savings of 2–10% are achieved; with strategic partnering, i.e. first generation partnering, savings of 30% are realistic, over time.' (Bennett and Jayes, 1995)

Figure 7.2► Partnering charter agreement: Railtrack/Extended Arm Contract

RAILTRACK

Extended Arm Contract (North West)

CHARTER

Our objectives are:

Safety

- Manage site safety to ensure no incidents, accidents or near misses
- Ensure compliance with legislation, Railway Group and Railtrack line standards
- Miminise Red Zone working

Teamwork

- Plan work in a timely and efficient manner in conjunction with Railtrack Zone and the train operating companies
- Achieve efficient and open communication
- Create a good and enjoyable working environment

Excellence

- Provide excellent engineering solutions
- Deliver the remitted work on time to the satisfaction of the Client
- Encourage innovation
- Make efficient use of available possessions
- Minimise use of disruptive possessions

Profit

- Make a profit
- Create value for money
- Achieve Project Delivery spend targets
- Provide accurate forecasts

23 May 1997

Birse

Source: Birse Rail/Railtrack; reproduced with permission

> 'A much more sophisticated second generation style of partnering, can deliver cost savings of up to 40%.' (Bennett and Jayes, 1998)

> 'Although still in its infancy, research suggests cost savings of 50% or more can be achieved by using third generation partnering.' (Bennett and Jayes, 1995)

In addition to these savings associated with the actual construction of the project, the costs associated with procurement are also likely to be reduced.

- **Improved quality and safety**: The existence of mutual objectives and the desire to make continual improvements in the design and construction processes are cornerstones of the partnering philosophy resulting in an improvement in quality and safety. Where strategic partnering is functioning, further benefits arise, including those derived from the additional learning and feedback mechanisms. These exist as a consequence of risk sharing and security of work.
- **Improvement in design and construction times and certainty of completion**: The efficiencies arising from the partnering arrangement previously discussed can lead to a large reduction in construction times. Also the attributes of the partnering arrangement, which include the benefits of team work and an element of risk and benefit sharing, lead to more certainty of completed project delivery dates.
- **More stable workloads and income**: A major difficulty faced by contractors is the lack of certainty about future workloads. This may have a negative impact on several aspects of the construction process, including the willingness of contractors to invest in the future in terms of recruitment, training and innovation. The use of strategic partnering provides an opportunity for forming long-term relationships and alleviates, to some extent, these difficulties. Contractors in a partnering arrangement are able to have knowledge about longer term expenditures on construction projects.
- **A better working environment**: Barlow *et al.* (1998) reported that several surveys have shown that a less adversarial atmosphere and shared commitment to projects result in perceived improvements in the working environment.

Concerns

It is necessary to acknowledge the existence of some important disadvantages to the use of partnering and other concerns regarding its use:

- **Initial costs**: There are additional costs associated with partnering. However, at less than 1% of the total project costs these are relatively small (Bennett and Jayes, 1995).
- **Complacency**: It has been suggested that a long-term partnering relationship that has the benefit of improved job security may lead to some complacency. Human nature is such that this could occur. However, there is possibly an equally strong counter-opinion that job security is highly valued and will therefore be safeguarded by continued effort and improvement. Some contractor participants have expressed the view that partnering, which is based upon trust, brings about a moral obligation which is more powerful than anything that may be contained in a traditional contract. In order to reduce this possibility of complacency, clients are able to use periodic competition and benchmarking to ensure that they are receiving the service expected. On this point, the perceived need for competition, described below, should also be considered.
- **Single source employment**: Strategic partnering could also result in a contracting organisation becoming very dependent upon one client and thus becoming extremely vulnerable, should this source of work be threatened. Clients may be similarly exposed, for example, where a single supplier or

contractor has developed a unique product or service. In this case, the problem can be avoided by ensuring the availability of other sources of supply.

- **Confidentiality**: Openness is a feature of partnering and this may cause some clients to be concerned about confidential matters. Disputes could occur if information, which may be regarded as commercially sensitive, was to be withheld from one or more of the partners.

> 'To avoid disputes as to disclosure, it is essential to identify as clearly as possible by category what information is to be shared. This will then encourage disclosure. It will also be important to seek to protect disclosed information by including a confidentiality clause in the partnering agreement.' (Long, 1998/99)

- **The perceived need for competition**: The accepted route to securing a good value price is through the competitive tendering process. The additional costs associated with an absence of competition are well understood and are used as a powerful argument against, for example, the use of negotiation. Also, accountability is a factor that is of concern for many clients. This may also be difficult to establish where there is a lack of competitive tenders. The traditional and easiest way of demonstrating value, is to show the results of a competitive tender selection process. Partnering may be judged to remove the competitive element, irrespective of the measures that may be taken to demonstrate otherwise, e.g. benchmarking.
- **Partnering through the supply chain**: In order to achieve the full benefits of partnering, the same principles need to be applied throughout the supply chain. While there is a great deal of positive feedback relating to partnering between clients and contractors, there appears to be less enthusiasm for this process among subcontracting organisations. There is evidence to suggest that some subcontractors believe that partnering has had a negative effect on their already dubious position and that conventional partnering does not work within the normal subcontractor framework (Smit, 1997).
- **Legal issues**: There is some doubt as to the legal status of the 'Partnering Charter'. McGeorge and Palmer (1997) have highlighted the concern that 'making explicit statements that a partnering charter does not create a legally binding relationship between partners does not necessarily mean that none exists'.

McGeorge and Palmer also make reference to a report from the Construction Industry Institute Australia (CIAA, 1996) which stated that:

> 'Although the construction contract provides a framework of rights and obligations, partnering has the potential to impact upon the allocation of risk established by the contract and subsidiary contracts. If the partnering arrangement breaks down, a party may find itself in a position where it is necessary, or at least attractive, to assert that the contractual risk allocation has been altered, either by the provisions of the partnering charter or by subsequent conduct or representations in the course of the partnering process. This is potentially the major risk to partnering in Australia.'

The authors also stress that they do not wish to infer that partnering is a legal minefield, but suggest that it is sensible to attempt to minimise any attendant risks, both legal and contractual. In a partnering agreement precautions can be taken by anticipating potential legal and contractual issues during the initiation stage of the partnering process.

This view is further supported in a Construction Law Bulletin published by Cameron McKenna (Long, 1998/99) which states that:

> 'The theory is that in the event that relationships deteriorated and the partnering arrangement collapsed, the parties should be able to rely upon their strict legal rights under the contracts. However, it is possible that, in implementing the partnering arrangement, the parties may be taken to have waived their strict legal rights under their contracts. The drawing up and signature of the charter could lay the basis for just such a waiver. Although it would be prudent to confirm in the charter that it is not intended to have any legal effect, this in itself may not be sufficient.'

This publication also highlights that there may be a need, in some situations, for a legally binding partnering agreement that, for example, could include details of the share of financial gains and losses relative to agreed targets. Where such a partnering agreement is to be used, the bulletin states that there is a requirement to ensure that the interface between the partnering agreement and the underlying contracts is consistent and workable. This includes reference to:

- **Payments**: In the event that a lump sum contract is used, the provision for payment under this form will need to be reviewed to ensure consistency with any incentive payment arrangement which may be part of the partnering agreement.
- **Liquidated damages**: The provision for liquidated damages within the underlying contract framework must be dovetailed with any related provision in the partnering agreement.
- **Quality**: The membership, roles and powers of the project management team, if set out or implied in a partnering arrangement, need to be exactly reflected by the underlying construction contract to avoid any confusion in respect of liability issues. Will the project management team be given power to instruct rework, replacement of defective materials and testing of materials and workmanship? Will this power remain with the consultant under the contract?'
- **Disputes**: The approach to the settlement of disputes is a central feature of partnering and, again, consistency is required between the partnering agreement and any provision in an underlying contract.

To a non-lawyer, there appears to be some doubt as to the legal effects of partnering, and clearly some caution is required where a partnering agreement is proposed. For this reason, it is suggested that readers undertake further research and seek further legal advice whenever the situation demands. The intention of this inclusion is merely to raise awareness.

7.6 Conclusions

The benefits attributed to partnering and the level of promotion it has been given and continues to be given at high levels suggest that it should have an important role to play in the future of any construction industry. However, it should also be recognised that significant negative opinion on the practice of partnering exists. A review of articles relating to partnering, published in the past few years, indicates the level of conflicting views. This suggests that it would be wise to be cautious in making any prediction about the direction of partnering and its future role.

It is also important to note that the extent of partnering activity, although seemingly rapidly growing, is a practice that remains yet untried by many clients. The results of a survey of 542 senior managers and directors commissioned by Galliford, a contracting organisation (Coulter, 1998), showed that nearly two-thirds of construction clients have never entered into a formal partnering arrangement with a contractor. The report also highlighted the rapid increase in the use of partnering in that, of those clients opting for partnering, two-thirds had done so since 1995.

The future of partnering is therefore unclear within the wider confines of the construction industry. The prevailing culture that underlies the construction industry contains strong elements of mistrust, cynicism and a general resistance to change. These cannot be easily overturned. Market realities should not be ignored, nor can the importance of the need for the delivery of a well-balanced education to future construction professionals. The industry is awaiting fully tried and tested methods of improvement that show only advantages and no disadvantages. This is not reality.

Culture and short-term thinking

The construction industry, more than most, is vulnerable to the uncertainties of the economy. In the midst of a construction boom (remember the late 1980s) contractors may have an abundance of work and be able to look to the future with great confidence. In such situations, the opportunities that partnering can offer may fade. Workload appears to be secure and margins are believed to be more than adequate. Similarly, and this is easier to recall, in the midst of a recession in the construction industry, clients can achieve what appears to be relatively staggering added value. In such circumstances, why not take the opportunity that the market offers and squeeze the last drop from the struggling contractor? Such short-term thinking will clearly hinder the progress of partnering.

The role of education

There is a need to recognise the importance of the education of future construction professionals in improving the overall performance of the construction industry. There is a need to relate real life experiences to students and, in so doing, prepare them for practical situations they may face. In so doing, it is perhaps too easy to perpetrate the defensive and isolated attitudes of the past.

It is necessary to prepare students for the real world. But it is unhelpful to ignore the need to change in the face of an under-performing construction industry measured against most criteria. Universities, therefore, have an important role to play in providing the industry with professionals who are able to accept fully the opportunity that partnering may provide.

References and bibliography

Barlow J., Cohen M., Jashapara A. and Simpson Y. (1998) *Partnering: Revealing the Realities in the Construction Industry*, Bristol Policy Press.

Bennet J. and Jayes S. (1998) *The Seven Pillars of Partnering; a Guide to Second Generation Partnering*, Reading Construction Forum Ltd; Thomas Telford Publishing.

Bennett J. (1999) *Partnering in Action*, SBIM Conference Proceedings.

Bennett J. and Jayes S. (1995) *Trusting the Team*, Centre for Strategic Studies in Construction, The University of Reading.

CIAA (1996) *Partnering: Models for Success*, Research Report No 8, Construction Industry Institute Australia.

Egan J. (1998) *Rethinking Construction*, Department of the Environment.

Coulter S. (1998) Most clients yet to try partnering. *Building*, June.

ECI (1997) *Partnering in the Public Sector: A Toolkit for the Implementation of Post-Award, Project Specific Partnering on Construction Projects*, European Construction Institute.

Latham, Sir M. (1994) *Constructing the Team*, HMSO.

Long P. (1998–99) Partnering agreements: the legal knot. *Construction Law Bulletin*, Cameron McKenna.

McGeorge D. and Palmer A. (1997) *Construction Management: New Directions*, Blackwell Science Ltd.

RICS (1995) *Improving Value for Money in Construction: Guidance for Chartered Surveyors and Clients*, University of Reading for Royal Institution of Chartered Surveyors.

Smit J. (1997) Chain reaction. *Procurement Supplement Building Magazine*; January.

Chapter 8

Benchmarking

8.1 Introduction

Benchmarking is a technique that, like other management practices discussed in this book, developed in the USA and is now successfully applied in many industry sectors around the world. It owes much of its relatively recent development to the success achieved by the Xerox Corporation. In the late 1970s, Rank Xerox were severely challenged by Japanese competitors who were able to manufacture photocopiers of a better quality and at a lower cost. In response, Xerox successfully implemented a programme of benchmarking which incorporated the comprehensive examination of their organisation throughout. Better practice, in any relevant process used by another company, was identified and translated into new Xerox practice. This facilitated a significant improvement in company performance and an improvement in their competitive position (Pickrell *et al.*, 1997).

Benchmarking can be applied across the business sector and throughout the entire hierarchy of a company from strategic to operational level. The aim of benchmarking is to improve the performance of an organisation by:

- Identifying best known practices relevant to the fulfilment of a company's mission. As will be discussed later, these may be found inside or outside the company.
- Utilising the information obtained from an analysis of best practice to design and expedite a programme of changes to improve company performance.
  It is important to stress that benchmarking is not about blindly copying the processes of another company, since in many situations this would fail. It requires an analysis of why performance is better elsewhere, and the translation of the resultant information into an action suited to the company under consideration.

In theory, therefore, benchmarking is a very simple process involving a clear pattern of thought:

Who does it better? →
How do they do it better? →
Adapt/adopt the better practices →
Improve performance

Despite the acknowledged benefits of benchmarking in assessing company performance and prompting its improvement, its use in practice within the UK appears to be low. Research carried out in 1997 by the Confederation of British Industry concluded that benchmarking is weak overall and that there was a need to raise awareness and practice of benchmarking across British industry (Confederation of British Industry, 1997). Any benchmarking practice at all, appears in the main to be restricted to large companies and in many instances is not carried out adequately. The level of use of benchmarking within the construction industry is consistent with this view. Some large construction organisations and construction clients have begun to use the tool successfully; however, the extent of use is very limited and in many companies is in its infancy.

The importance of benchmarking to the construction industry has recently been acclaimed at a very high level. In *Rethinking Construction*, the report of the Construction Task Force (Egan, 1998) to the Deputy Prime Minister, John Prescott, it was stated that:

> 'Benchmarking is a management tool which can help construction firms to understand how their performance measures up to their competitors' and drive improvement up to "world class standards" . . .'

The report makes clear the under-performance of the construction industry and the need for fairly dramatic improvement, much of which is considered to be reliant upon successful benchmarking practice. It also highlighted the significance of the construction industry to the British economy which, in 1998, had an output of £58 billion and a workforce of approximately 1.4 million, further emphasising the need for change.

In other sections of this book, recognition has been given to the poor rate of implementation of new management practices. This introduction suggests that benchmarking practice in the UK may also currently be seen to fall into this group of fashionable but little used activities. However, in the case of benchmarking, business survival may be dependent upon it and, therefore, its wide acceptance may become a natural consequence. It is likely that the use of benchmarking practice will accelerate, and as companies improve by it, the need for other organisations to follow will be apparent.

8.2 The need for benchmarking

It is clear from the above that the purpose of benchmarking is to improve a company's performance with the aim of maintaining or achieving a competitive edge. The UK construction industry is witnessing a rapid change in many areas, some of which are of great significance. These include, for example, methods of procurement, client expectations, client and contractor relationships, professional appointments, the Private Finance Initiative, government focus, warranties and guarantees, payment provisions and safety regulations. Consideration of this incomplete list should otherwise convince anyone who believes that the construction industry stands still. In addition, there is an increasing level of competition from Europe and elsewhere, including the USA and Japan, where benchmarking practice is now part of the culture. There are good incentives for the use of benchmarking. The advantages that it may bring include:

- An opportunity to avoid complacency which may be nurtured by monopolistic situations, resulting in the loss of market leadership.
- An improvement in the efficiency of a company, leading to an increase in profits.
- An improvement in client satisfaction levels, resulting in an improved reputation, leading to more work, better work and an increase in profits.
- An improvement in employee satisfaction in working in a more efficient, profitable and rewarding environment.

Many organisations that have successfully used benchmarking in order to achieve a specific objective continue to use it thereafter as part of a total quality management programme. In this way it can contribute to the long-term objectives of the company which primarily may be to achieve a high level of customer satisfaction, recognised excellence in their sector and, thereby, a high return.

8.3 Benchmarking practice

In practice, benchmarking can be very demanding in terms of the commitment required to carry it out, if not in direct costs, then in human resources. It is therefore important to ensure that, prior to commencement of a study, everything necessary for its success is in place. At the outset, the following key decisions and actions are suggested:

- Designation of someone within the organisation to take on the role of project leader.
- Establishment of clear project objectives.
- Establishment of an achievable programme.
- Determination of the need for a project team, its composition and any training required.
- Sanctioning of any resources required.

Once this preliminary stage is complete, the benchmarking exercise can be carried out in the confidence that there are clear objectives, a reasonable time frame in which to work, adequate resources, and a project team that is ably led.

Benchmarking framework

There are several variables which may impact upon a benchmarking study, including, for example, deadlines for completion, information availability, benchmarking objectives and management resources. Each benchmarking exercise will thus require a flexible approach. The suggested outline that follows is therefore provided with this consideration in mind.

Sequence of benchmarking activities

- Understand the processes within the organisation and identify the factors which are critical to its success.
- Determine what you wish to benchmark.
- Establish the sources of benchmarking data.
- Collect the data.
- Analyse the data to identify best practice.
- Examine the best practice to determine how it is achieved.
- Prepare and implement an action plan.

This logical and practical framework will provide information relating to better practice in other companies or, if the study is internal, in other parts of the same company. Once established, this can be used as a basis for implementing changes within an organisation with the intention of matching or improving the performance identified as best practice. This outline is further discussed below.

Understanding the processes

It is probable that most senior staff within an organisation believe they have a clear and detailed understanding of what their company does. While this is possible in general terms, the intricacies of most organisations within the construction industry are such that this knowledge is likely to be superficial in many areas. Therefore, since benchmarking requires a detailed appreciation of the activities of a company, an essential first step is the thorough examination of an organisation in order to clarify its processes and their interrelationships. There are several possible approaches and techniques that can be used to assist with this requirement. Workshops, incorporating a variety of tools and techniques, such as, brainstorming, mind mapping, process mapping, fishbone diagramming, etc., can be used at a strategic level at which to produce an outline of an organisation's processes. This might be the construction process or a detailed analysis of a single process, such as, procuring subcontract works and further subprocesses, for example, preparing subcontract packages.
The fishbone diagram (Figure 8.1) illustrates a process analysis for *procuring subcontract works* and indicates the hierarchical relationships between processes and subprocesses occurring in differing layers of an organisation and how activities may be related to a company's mission. The product of this type of workshop exercise is the identification and understanding of the core processes within an organisation.

What to benchmark?

It is possible to benchmark a multitude of processes and this surfeit of opportunity is a problem that needs to be overcome in the initial stages of a study. To improve effectiveness, the benchmarking exercise should focus on relatively few processes and incorporate those that are most likely to provide an early payback. Therefore, as a first stage, a list of key processes must be established with a view towards obtaining the most achievable and valuable benchmarking output. The process-mapping techniques discussed above will lead to an

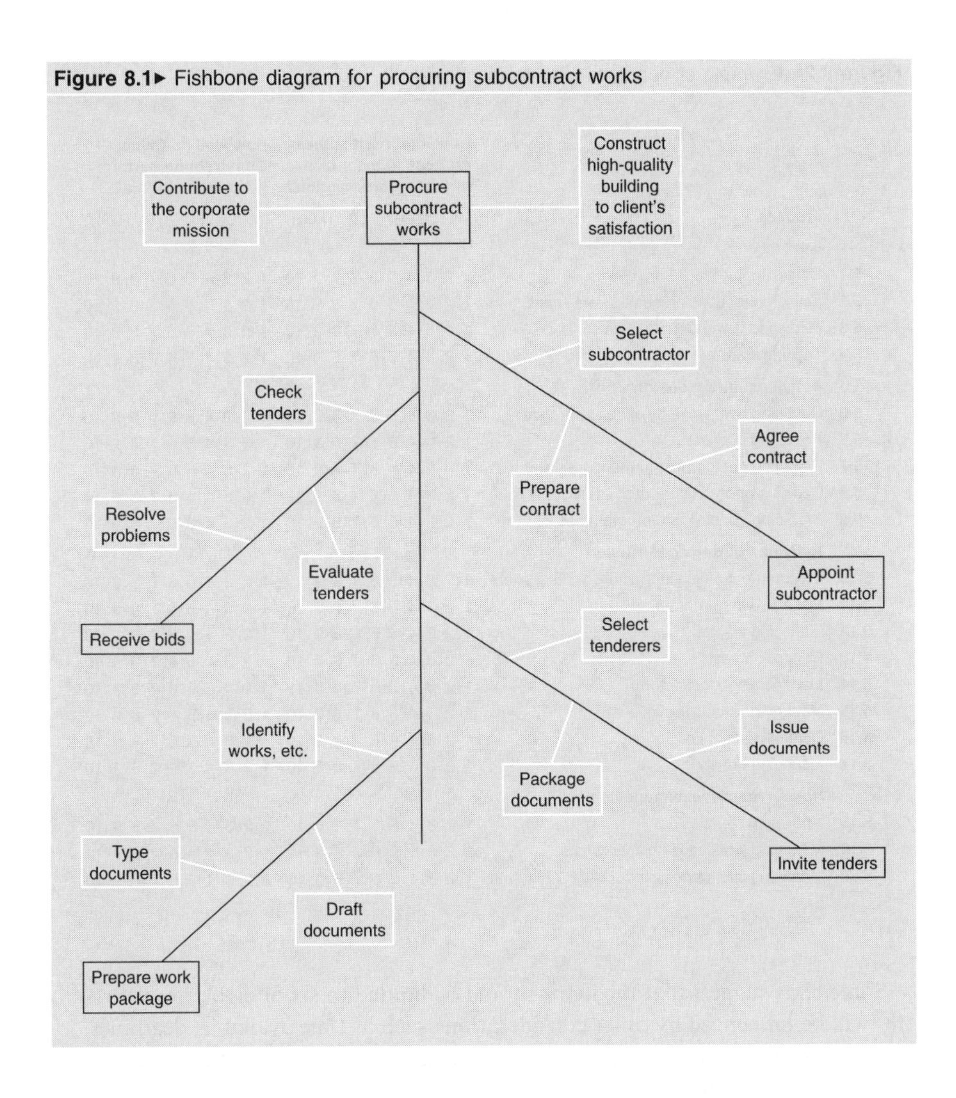

Figure 8.1▸ Fishbone diagram for procuring subcontract works

appreciation of the value added by each of the processes and subprocesses and will assist in the identification of factors critical to the success of an organisation. To further assist in this task, valued and meaningful opinion of the organisation should be sought from anyone with a direct association, including staff, clients, subcontractors and consultants. A well-tested method of obtaining useful information is by use of a questionnaire that may provide valuable qualitative feedback. Structured interviews may also be carried out with clients, and may afford more meaningful feedback than that obtained via questionnaires. An example of a questionnaire used in a recent benchmarking exercise is shown in Figure 8.2. Respondents were asked to indicate their opinion as to both importance and performance.

A survey such as this acts as an awareness exercise and assists in the identification of those processes that are perceived as high importance or low performance, and will thus be likely to benefit from the benchmarking process.

Figure 8.2▸ Example of benchmarking questionnaire

	Process	How important is the process to the success of Civco Construction? (V. LOW) (V. HIGH)	How well do Civco Construction perform with regard to this? (V. LOW) (V. HIGH)
	General		
5.1	Obtaining feedback from clients	1 2 3 4 5 6 7 8 9 10	1 2 3 4 5 6 7 8 9 10
5.2	Acting on feedback received from clients	1 2 3 4 5 6 7 8 9 10	1 2 3 4 5 6 7 8 9 10
5.3	Marketing Civco Construction	1 2 3 4 5 6 7 8 9 10	1 2 3 4 5 6 7 8 9 10
5.4	Training staff	1 2 3 4 5 6 7 8 9 10	1 2 3 4 5 6 7 8 9 10
	Activities during design		
5.5	Obtaining best value by option appraisal	1 2 3 4 5 6 7 8 9 10	1 2 3 4 5 6 7 8 9 10
5.6	Preparing budgets	1 2 3 4 5 6 7 8 9 10	1 2 3 4 5 6 7 8 9 10
5.7	Preparing design/procurement programmes	1 2 3 4 5 6 7 8 9 10	1 2 3 4 5 6 7 8 9 10
5.8	Deciding upon procurement strategy	1 2 3 4 5 6 7 8 9 10	1 2 3 4 5 6 7 8 9 10
5.9	Preparing construction programmes	1 2 3 4 5 6 7 8 9 10	1 2 3 4 5 6 7 8 9 10
	Activities during construction		
5.10	Procurement of subcontractors and suppliers	1 2 3 4 5 6 7 8 9 10	1 2 3 4 5 6 7 8 9 10
5.11	Managing the programme	1 2 3 4 5 6 7 8 9 10	1 2 3 4 5 6 7 8 9 10
5.12	Managing waste	1 2 3 4 5 6 7 8 9 10	1 2 3 4 5 6 7 8 9 10
5.13	Controlling costs	1 2 3 4 5 6 7 8 9 10	1 2 3 4 5 6 7 8 9 10
5.14	Managing subcontractors	1 2 3 4 5 6 7 8 9 10	1 2 3 4 5 6 7 8 9 10
5.15	Managing sites to improve safety	1 2 3 4 5 6 7 8 9 10	1 2 3 4 5 6 7 8 9 10
5.16	Controlling finance	1 2 3 4 5 6 7 8 9 10	1 2 3 4 5 6 7 8 9 10
5.17	Managing quality	1 2 3 4 5 6 7 8 9 10	1 2 3 4 5 6 7 8 9 10
	Human resource management		
5.18	Motivating staff	1 2 3 4 5 6 7 8 9 10	1 2 3 4 5 6 7 8 9 10
5.19	Managing workloads and resources	1 2 3 4 5 6 7 8 9 10	1 2 3 4 5 6 7 8 9 10
5.20	Managing staff performance	1 2 3 4 5 6 7 8 9 10	1 2 3 4 5 6 7 8 9 10

Some texts suggest that the items should be limited to six or eight; however, this will be influenced by other considerations such as time available, deadlines, the availability and reliability of data sources and whether or not the exercise is to be internal or external.

8.4 Construction industry key performance indicators

A recent initiative introduced by the government-appointed task force, *Movement for Innovation*, is the introduction of an intended series of ten key performance indicators (Table 8.1) which may provide a further means of identifying potential performance gaps.

The value of the industry averages in Table 8.1 is questionable at the present time since they represent a starting point in the initiative, are based upon limited information and rely only on limited research. Nonetheless such key performance indicators could, in future, provide contractors with useful measures for making industry-wide comparisons. However, the use of the information by clients in preparing short lists for tendering purposes raises some doubts. It should be

Table 8.1► Ten key performance indicators

	Key performance indicator (KPI)	Industry average	How KPIs are measured
	Project		
1	Client satisfaction: product	8	On a scale of 1 (totally dissatisfied) to 10 (satisfied)
2	Client satisfaction: service	8	On a scale of 1 (totally dissatisfied) to 10 (satisfied)
3	Predictability of cost		
	Design cost	0%	Difference between design cost at decision to go ahead and at end of construction
	Construction cost	1%	Difference between estimated cost at start and final cost
4	Predictability of time		
	Design time	18%	Estimated design time compared with actual time
	Construction time	9%	Estimated construction time compared with actual time
5	Defects	3	Measured on a scale of 1–4 (with 4 representing a project free from defects)
6	Construction cost	–3%	Inflation-adjusted improvement for the year-on-year cost of a given project
7	Construction time	–1%	Year-on-year comparison of improvements in delivery time
	Company		
8	Profitability	3.2%	Pre-tax profit as a percentage of sales
9	Productivity	£60,000	Turnover per full-time employee
10	Safety	997	Reportable accidents per 100,000 employees per annum

Source: Smith K. (1999)

borne in mind that the data are based, in part, upon subjective viewpoints and are reliant on accuracy and consistency in collection. This may be difficult to monitor adequately in practice. The value of the key performance indicators is more likely to be found in providing a structure with which companies can establish their own baselines that may be used to monitor their own future performance.

8.5 Sources of benchmarking data

Benchmarking may be performed by access to several reference points. In the case of a large national contracting organisation, a multi-tiered approach may be possible, using, for example, internal comparisons such as:

- Comparisons within a section, region or division (e.g. Civco Homes; Civco South West).
- Comparisons within the entire company, including all regions and divisions (e.g. Civco Holding).

and external comparisons, such as:

- Comparisons with similar competitors in the UK construction industry.
- Comparisons with non-competitors, which might include those outside the construction sector.

The choice of level of data will be determined by several factors, which include:

- Objectives of the study.
- Need for confidentiality.
- Co-operation of third parties.
- Speed in which the completed study is required.
- Data availability.
- Resource availability.

In practice, the question '*Who to benchmark against?*' may be answered initially by the desire for a simplified process. This will eliminate the problem of finding suitable external partners and provide easier access to commonly understood and presented data. The limitations of an internal approach are clear: best practice within an organisation is unlikely to be best practice within an industry. In the event that a study seeks to involve external partners, it should be remembered that, although the construction industry has unique features, many of the processes used are common to all businesses. Value can therefore be gained by identifying best practices beyond the confines of the construction sector alone. Reference to published industry key performance indicators, discussed above, does allow for external comparisons, however, the benchmarking process cannot be extended beyond this unless information about best practices are available and can be analysed properly and effectively.

Obtain the data

Once the sources of data are established, the types of information and appropriate methods of retrieval need to be determined. These include:

- **Qualitative data** – e.g. opinion of the performance of a company – may be sought by the use of questionnaires or structured interviews. Detailed guidance on the use of these techniques is provided in other texts. In deciding upon which approach to adopt, consideration should be given to the level of information required and the situation of the respondents. Questionnaires, for example, may be an appropriate method of identifying staff perceptions of company performance where a reasonable response can be expected (see above), but structured interviews will promote more open feedback and may be a better means of gathering information from senior representatives of client organisations.
- **Quantitative data** – e.g. value of design changes as a proportion of final cost – can be used to measure or gauge performance. While it is likely that qualitative data will be relatively inexpensive to obtain, quantitative data may require greater resources. For example, in a recent study in which the author was involved, quantitative data were obtained by comprehensive data sheets for approximately 200 construction projects to identify relative performance and

improvements over time. The information gathered incorporated that relevant to project costs, design times, change orders, wastage, defects, site management, subcontract performance and design performance. The data were collected manually by internal personnel and stored in a database; a time-consuming activity.

Analyse the data

As with any research project, a consistent and rigorous analysis of data is necessary. Methods of analysing data may be complex and are considered to be beyond the scope of this book. Some training or support may be required in both statistical analysis and software application to allow the execution of an adequate analysis.

Implement identified improvements

The main objective of benchmarking is to improve performance, not merely the identification of a performance gap. When opportunities for improvement are clear, a strategy can then be determined by the benchmarking team to accommodate the improvements necessary to close the apparent performance gap.

8.6 Factors critical to the success of benchmarking

- *The support of senior management*. Token support will not be enough. It is probable that, in most organisations, benchmarking will be carried out without relief from other responsibilities and in this sense is an added burden. A great commitment to benchmarking through ownership, active involvement and leadership will therefore be required.
- *An understanding of the processes of an organisation.*
- *Adequate resources*. Although a study may be carried out internally, without the use of an external consultant, staff time (often highly paid staff time) is likely to be an essential ingredient.
- *Staff development*. Staff participating in a benchmarking study should be familiar with methods used to collect and analyse the data that are necessary to identify and examine best practice.
- *Team participation.*
- *Adequate planning of the study*. A timetable of the study and subsequent action plan should be established and monitored.

8.7 What can benchmarking bring to the construction industry?

The low standing of the construction industry within the UK suggests that there is significant opportunity for benchmarking to be used with great effect. As discussed in other chapters of this book, clients too frequently criticise the industry in that, buildings are delivered late, exceed budget costs and are of poor quality. This provides three broad areas for improvement. This view has recently been enforced by the publication of the report, *Rethinking Construction* (Egan, 1998) which stated that:

> 'In construction, the need to improve is clear. Clients need better value from their projects, and construction companies need reasonable profits to assure their long-term future. Both points of view increasingly recognise that not only is there plenty of scope to improve, but they also have a powerful mutual interest in doing so.'

The inefficiencies and waste in construction are demonstrated by the following examples of poor performance, which have been identified by recent studies in the USA, Scandinavia and the UK (Egan 1998):

- As much as 30% of construction work is carried out to remedy problems of first-time construction.
- The output of the labour force is at 40–60% of optimum.
- Poor site safety is demonstrated by an accident rate which can result in 3–6% of total project costs.
- Material waste is high, with wastage being 10% or more.

As stated above there may be difficulties to overcome in carrying out benchmarking studies. One of the major problems relates to the accessibility of information that may prove difficult to obtain in all studies but particularly where external sources are being used. Benchmarking organisations have been established to assist with this; the recent proposal in the Egan report suggests that an information centre should be set up whereby those involved in the construction industry can have access to a bank of data that will prove helpful in benchmarking exercises.

Benchmarking and partnering

To justify the adoption of partnering, clients may need to validate the benefits of such an arrangement by establishing that good value and continued performance improvements exist. Since the usual assurance of value, traditionally dependent upon the receipt of competitive bids, is absent, benchmarking may be required to provide an alternative method of measuring the relative performance of a partner contractor. This would then help to demonstrate the added value arising from a partnering arrangement.

Benchmarking and the consultant

Throughout this chapter, benchmarking has been viewed in the main from the perspective of the construction company and members of its supply chain. This is a natural inclination since construction may, to some extent, be recognised as a manufacturing process with an easily recognised and evaluated product, i.e. a building. It would be mistaken, however, to ignore the opportunity that benchmarking may offer to construction consultants who also operate within the same increasingly competitive and changing domain. For instance, Pickrell *et al.* (1997) present a case study of the Bucknall Group which describes how the company introduced benchmarking to combat the difficulties it faced in the early 1990s. One example of the success achieved related to the introduction of an electronic time sheet system that fed into the costing system and saved an estimated equivalent of £55,000 per annum. In addition to improving the profitability of a practice, benchmarking can also be used by clients to compare consultant performance in a similar fashion to that discussed above for contracting organisations.

References and bibliography

Bennett J. and Jayes S. (1995) *Trusting the Team*, Centre for Strategic Studies in Construction, The University of Reading.

Confederation of British Industry (1997) *Benchmarking the Supply Chain*, Partnership Sourcing Ltd.

Egan J. (1998) *Rethinking Construction*, Department of the Environment, Transport and the Regions.

Cook S. (1997) *Practical Benchmarking. A Manager's Guide to Creating Competitive Advantage*, Kogan Page.

Karlöf B. and Östblom S. (1993) *Benchmarking: A Signpost to Excellence in Quality and Productivity*, John Wiley & Sons.

Latham, Sir M. (1994) *Constructing the Team*, HMSO.

McGeorge D. and Palmer A. (1997) *Construction Management: New Directions*, Blackwell Science Ltd.

Pickrell S., Garnett N. and Baldwin J. (1997) *Measuring Up: a Practical Guide to Benchmarking in Construction*, Construction Research Communications Ltd (by permission of the Building Research Establishment).

RICS (1995) *Improving Value for Money in Construction, Guidance for Chartered Surveyors and Clients*, University of Reading for Royal Institution of Chartered Surveyors.

Smith K. (1999) How do you shape up? *Construction News*, 20 May, pp. 12–13.

Chapter 9

Procurement

9.1 Introduction

A client, who has made the major decision to build, is faced with the task of procuring the construction works that are required. This may be a daunting prospect, given the level of financial commitment and other risks associated with the venture, the complex nature of construction and the possible perception of the construction industry as one that frequently under-performs in its work. The wide range of procurement systems now available, and the understanding that procurement choice may have a significant bearing upon the outcome of a project, signifies both the opportunity and importance of meeting the procurement challenge with a well-considered strategy.

In deciding upon a procurement strategy, recognition should be given to client priorities and, where possible, desires, in terms of project duration, cost and performance objectives. In order to determine an appropriate method of procurement, it is necessary to match the needs of the client with the most suitable procurement approach available. To enable this, an understanding of both the client's objectives and the operation and relative attributes of the available procurement types is required. This chapter looks at the range of procurement methods available and considers how the procurement strategy can be selected to meet client needs. Procurement is a decision process that provides substantial added value for the client.

9.2 Procurement trends

Traditional procurement systems remain the most frequently used in practice. However, in recent years, there has been a significant shift towards alternative strategies. Tables 9.1 and 9.2 indicate the trends in procurement methods used between 1984 and 1995.

Table 9.1▸ Trends in methods of procurement: by value of contracts

Procurement method	Contracts (% of total, by year)						
	1984	1985	1987	1989	1991	1993	1995
Lump sum: firm BQ	58.7	59.3	52.1	52.3	48.3	41.6	43.7
Lump sum: spec. and drawings	13.1	10.2	17.7	10.2	7.0	8.3	12.2
Lump sum: design and build	5.1	8.0	12.2	10.9	14.8	35.7	30.1
Remeasurement: approximate BQ	6.6	5.4	3.4	3.6	2.5	4.1	2.4
Prime cost plus fixed fee	4.5	2.7	5.2	1.1	0.1	0.2	0.5
Management contract	12.0	14.4	9.4	15.0	7.9	6.2	6.9
Construction management	–	–	–	6.9	19.4	3.9	4.2

Source: Davis, Langdon & Everest (1996)

Table 9.2▸ Trends in methods of procurement: by number of contracts

Procurement method	Contracts (% of total, by year)						
	1984	1985	1987	1989	1991	1993	1995
Lump sum: firm BQ	34.6	42.8	35.6	39.7	29.0	34.5	39.2
Lump sum: spec. and drawings	55.7	47.1	55.4	49.7	59.2	45.6	43.7
Lump sum: design and build	2.4	3.6	3.6	5.2	9.1	16.0	11.8
Remeasurement: approximate BQ	3.2	2.7	1.9	2.9	1.5	2.3	2.1
Prime cost plus fixed fee	2.3	2.1	2.3	0.9	0.2	0.3	0.7
Management contract	1.8	1.7	1.2	1.4	0.8	0.9	1.2
Construction management	–	–	–	0.2	0.2	0.4	1.3

Source: Davis, Langdon & Everest (1996)

The data in Tables 9.1 and 9.2 indicate that:

- The use of firm bills of quantities is the main method of procurement (by value) and the apparent decline in this form of procurement has now steadied.
- The growth of design-and-build procurement appears to have stopped. This now accounts for approximately 30% of contracts let by value and 12% by number. The increased use of this method of procurement since 1984 is dramatic.
- The use of construction management and management contracting, after rising to approximately 27% by value of contracts in use in 1991, has declined and appears to have settled at 10–11%.
- The relatively low usage of both remeasurement and prime cost contracts in 1984 (approximately 11%) has further declined, accounting in 1995 for less than 3% of contracts in use by value.

The publication of the results of a more up to date *Contracts in Use* survey is imminent at the time of publication.

9.3 The needs of the client

Clients may have a major influence on a project throughout its development and construction and may be actively involved in contributing to and monitoring its progress at each stage. Research has shown (Masterman, 1992) that clients rate very highly the need to be both informed and involved. The full involvement of the client at briefing stage is of particular importance since it is a key aspect of good briefing, which is a fundamental requirement of a successful project. As in most businesses, the construction industry must meet the needs of a widely ranging client body with varying degrees of knowledge, experience, resources and objectives. The ability and willingness of clients to contribute will be dependent upon these factors and also upon individual personality, which, in the author's experience, can play a major role.

Client categories

Procurement strategy should be considered in light of the needs and characteristics of the client, and an understanding of client types and their priorities is therefore important. Although each client will be unique, the following classifications, which have been widely accepted, allow us to categorise clients by key characteristics (Masterman 1992):

- **Public or private sector**: The priorities of these groupings will generally differ in terms of the balance of time cost and quality objectives, accountability and certainty of output. For example, public sector clients are more likely than private sector clients to be driven by the need for low cost, cost certainty and accountability.
- **Experienced or inexperienced**: It is difficult to provide a precise definition of an experienced or inexperienced client although it is important to appreciate the difference between these. It is reasonable to assume that the latter will need considerably more advice, more support and will be less able to comply with the needs of management-based procurement methods. Experienced clients may have a working knowledge of the construction industry, may retain the services of in-house construction advisers and will be able to contribute throughout the process. Without these attributes, inexperienced clients may have unrealistic expectations and the tendency to inappropriately interfere rather than contribute.
- **Primary constructors or secondary constructors**: The main business of primary constructors is the construction of buildings for sale or lease, e.g. speculative developers. As inferred by the label, secondary constructors seek to build to provide facilities to support their main business activities.

With reference to the above classifications, the main categories of client that are likely to be encountered in practice include:

- Public, experienced, primary constructors (e.g. local authorities).
- Private, experienced, primary constructors (e.g. commercial developers).
- Private, experienced, secondary constructors (e.g. major corporations with regular development needs).

- Private, inexperienced, secondary constructors (e.g. small manufacturing organisation building a new premises).

It is recognised that these different client types will have different procurement needs. In so doing, it should be borne in mind that generalisations may prove to be inaccurate when considering the situation of a specific client. It would be wrong, therefore, to automatically label clients in accordance with the above classifications and assume their objectives to be fully compliant with a client type.

Client procurement needs

The clients' key project objectives and their interrelationships should be understood before an appropriate procurement strategy is selected. The criteria that need to be considered include:

- **Time**: Project duration or completion dates may be critical, e.g. the Millennium Dome, to the success of a project and in some situations, if not met, could lead to total failure in meeting a client's objectives. While most clients are likely to have a desire for an early building completion, it is important to distinguish between this and true need since attempting to meet the objective of early completion is likely to have consequences on other project requirements. The choice of procurement strategy can have a significant bearing on the total duration of a project and this factor is considered further below.
- **Cost**: In the event that a limited capital budget is the prime consideration of the client, quality, in the form of a reduced specification, is likely to be restricted and the project duration will be the optimum in terms of construction cost rather than client choice. It is important to mention again (see Chapter 6 on Value management) the importance of differentiating between cost and value. If budget is the prime objective of a client, a resultant reduction in specification levels and perceived building quality could lead to a relatively greater reduction in project value.
- **Quality**: The quality of a building is influenced by several factors, including: the briefing process, the suitability of materials, components and systems, and their interrelationships within the total design. The quality control procedures that are in place during both design and construction will be paramount. The choice of procurement strategy can affect the design process and means of control by which clients and their advisers can monitor both specification and construction activity. It should be noted that quality is a partially subjective issue, and is sometimes difficult to define and identify. It may not necessarily mean a more complex building or a higher specification.
- **Accountability**: Organisations in receipt of public funding will naturally be concerned with this aspect since they are subject to public scrutiny. It is sometimes assumed that accountability is less of a concern for companies in the private sector. This assumption may be misplaced. Research carried out by Masterman in 1988 showed that 'private, experienced secondary clients – i.e. major and active manufacturers, retailers, service organisations, etc.' indicated that accountability is the most important criterion in ensuring project success (Masterman, 1999).

Figure 9.1▶ Matrix of time, cost and quality

Priority	Dependent criteria: Time	Cost	Quality
Time		↑	↓
Cost	↑		↓
Quality	↑	↑	

Relationship of key criteria

- If the main client priority is time, costs may rise and quality may suffer.
- If the main priority is low cost, time to obtain may increase and quality may suffer.
- If the main emphasis is quality, time to obtain may increase and costs may rise.

- **Cost certainty**: The degree to which cost certainty is required, prior to commitment to construction or at project completion, restricts procurement choice considerably. The risks associated with abortive design fees are also issues that clients should be aware of.
- **Availability of expertise**: It is probable that large experienced client organisations will have in-house expertise that would permit the consideration of a managed approach to construction. Smaller occasional clients are unlikely to be in this position.
- **Certainty of project objectives**: Some forms of procurement, such as management contracting, incorporate an inherent facility to accommodate design development throughout a project. Others are particularly unsuited to design changes, e.g. design and build.

The relative strengths and weaknesses of the available procurement options may be evaluated in terms of the above criteria. Since each procurement strategy will contain a differing balance of the various attributes required by clients, a prioritisation of key objectives is necessary to enable the most suitable choice to be made. In deciding upon a procurement strategy, it is important to appreciate the interdependencies of the selection criteria and understand that an emphasis on one aspect will impact upon one or more of the others. The matrix shown in Figure 9.1 indicates how the key criteria of time, cost and quality may interrelate in some situations.

Please note that, for simplicity, this model excludes other procurement selection criteria, that have been outlined above, and is based upon hypotheses that may not always apply.

9.4 Procurement options

Several procurement options are available to the client, and within each broad type there are several variants, each of which may be possibly refined to accommodate particular client needs and project specifics. For example, within a traditional arrangement, it is normal to have some of the works carried out under a cost-plus or remeasurement arrangement and it is also possible to let a portion of the works on a design-and-build basis. An appreciation of the operation and application of each of the procurement options is essential to developing a sound procurement strategy.

Traditional

The key feature of this form of procurement is the separation of design and construction. The client appoints a team of consultants – which was, in the past, frequently led by the architect – to design the building and prepare tender documentation. Competitive tenders are invited for the construction works which are subsequently let to a single contractor who is perceived as providing the best offer in terms of a balanced view of risks and cost. The main advantages and disadvantages of this procurement option are as follows.

Advantages

- Provided that the design process has been completed fully in the pre-contract stage, and this condition is regularly unfulfilled, a high level of price certainty for the client is achieved since cost is known before construction commences. Price certainty will be affected by the manner in which the client and the team decide to deal with building cost inflation in the post-contract stage, although, in present times of low inflation rates, this may seem a relatively minor issue. Price risk is increased if contracts provide for the reimbursement to contractors of building cost escalations.
- Opportunity to control the appointment and activities of the designers.
- Assuming the existence of fair competition, a low tender price.
- It accommodates design changes and aids the cost management process (subject to the basis of tender submission).
- Accountability.
- Where bills of quantities are used, a reduction in tender preparation costs and an improvement in tender quality.

Disadvantages

- The time from inception to start on site will be relatively lengthy – an outcome of the need to prepare full documentation prior to commencement of construction.
- While price certainty is cited above as an advantage, in practice this is often much more elusive. Post-contract design changes are frequently abundant and resultant delays and disputes are common. This position may be exacerbated by the separation of design and construction responsibility. The constructor may have less incentive to mitigate the effects of design problems and, in some cases, may find opportunity to gain financially when they occur.
- The constructor's perspective, which could enhance buildability, is likely to be absent from the design.

Design and build

With this method of procurement, the contractor accepts the risk for the design element of a project. It is common for the client, initially, to appoint design consultants to develop a brief, examine feasibility and prepare tender documents that will include a set of Employer's Requirements. Contractors are invited to tender on the basis that they will be responsible for designing and constructing the project and will submit a bid which will incorporate design and price information. The Contractor's Proposals will be examined by the client and the project subsequently let. An issue to be considered in the early stages of the project is the nature of the Employer's Requirements, which may vary significantly in terms of detail. Clients may need to balance their conflicting desires to both direct the design and transfer full design risk to the contractor. *Develop and construct* is an approach whereby the client's consultants prepare a concept design and ask contractors to develop that design and construct the works.

The key advantages of using design and build are as follows.

- **Single-point responsibility**: Many of the problems arising from the use of traditional procurement relate to the division of design and construction responsibility. Contractor's claims for additional monies and extensions of time relating to design matters are common. With design-and-build procurement, it is reasonable to assume that contractors will be motivated towards the reduction of design problems and their mitigation in the event that they arise. Although, elsewhere in this book (see Chapter 10, Risk Management) the possible negative impact on quality, where design problems arise in design-and-build procurement, is discussed.
- **Price certainty prior to construction**: With design and build, provided there are no client changes, a high level of price certainty exists. However, in practice clients are often reluctant to forgo this consideration. The risk relating to design fees prior to commitment to construct is also reduced. Design advisers employed independently by the client will be appointed on a significantly reduced service and fee basis.
- **Reduced project duration**: This is made possible due to the overlap of the design and construction phases. The increased risks associated with this, in terms of uncertainty of total design, cost and time implications at time of commitment, are borne by the contractor. It is reasonable to assume that this increased risk will be reflected directly or indirectly in the cost of construction.

In addition to these advantages, there are other possible benefits, although these are less certain. An improved degree of buildability may be achieved since the contractor has a greater opportunity to influence the design. However, in practice this will be dependent upon the level of design direction contained in the Employer's Requirements. For example, develop and construct offers less opportunity than a turnkey system. The proximity and relationship of the contractor to the design team is also important. Also, with design and build, there is the potential to extend design liability beyond *reasonable skill and care* to include *fitness for purpose*. However, the non-availability of insurance for the increased design liability and the possible questionable ability of contractors to withstand the impact of a large claim in the event of failing to achieve the fitness-for-purpose requirement are such that it will normally be preferable to forgo this opportunity (Morledge and Sharif, 1996).

There are several disadvantages to the use of design-and-build procurement. These include:

- **Client's reduced ability to control design**: As discussed above, the scale of this concern is dependent upon the design-and-build approach that is adopted. Clients may wish to substantially develop a design before transfer to the contractor and also to impose the client design team upon the contractor. In any event, during the post-contract stage, the design team are employed to serve the needs of the contractor and thus client control is diminished. This position may be confused where the client's designers are imposed upon the contractor and could understandably be expected to show continued loyalty to their client.
- **Commitment prior to full design**: The nature of design and build is such that complete design details will not be available until some time after the contractor is appointed. In essence, therefore, the client contracts to buy a building that is yet to be fully designed. It is likely, in the author's experience, that the client will be dissatisfied with an aspect of the design as it develops and, provided the contractor's design is in accordance with the contract, be obliged to issue change orders, adding to the contract sum.
- **Difficulty in comparison of tenders**: The evaluation of the differing design alternatives contained within the contractor's proposals adds a significant complexity to the normal tender review process. Design quality is difficult to quantify and consensus within client organisations and design teams may be difficult to obtain.
- **Cost management difficulties**: It is unlikely that there will be any adequate price information available to the client for the evaluation of contract changes. This may create significant cost management problems.

Management-based strategies

There are several possible variants to a management-based procurement strategy. Each shares the main characteristic of the appointment of a party, usually a contractor, to manage the *construction* works and, in the case of design and management, also the *design* of the works, which is done in return for a lump sum or percentage fee. The scope of the management provision and methods of establishing contractual links in the construction supply chain provide differentiation between specific types. This approach to procurement is particularly suited to large, complex projects whereby it may be beneficial to reduce risk for the contractor (Murdoch and Hughes, 1992).

The roles and responsibilities assumed by the parties involved in the project are somewhat different to those in a traditional procurement arrangement. The position of the management contractor is that of a consultant who takes on the role of manager of the process and does not carry out any of the construction works directly, other than the provision of preliminary type items, e.g. site accommodation, items of plant, etc., where required. The actual construction is performed by works contractors who, instead of being cast as one of the many anonymous subcontractors within the traditional framework – excluding those elevated by the action of nomination – achieve a much loftier position, perhaps more accurately reflecting their real contribution to the construction process.

Management contracting

This procurement option involves the appointment of a management contractor who is paid a fee for procuring and managing the construction works. The works are let on a work package basis, usually by competitive tender, although other methods of procuring individual works packages may be adopted. Contracts are arranged between the management contractor and the works contractors.

The key advantages of using management contracting include:

- **Early involvement of the management contractor**: This promotes an early construction-based perspective, enhancing buildability and project planning.
- **Reduced project duration**: Construction work may start as soon as sufficient work has been designed, resulting in overall time benefits due to the overlap of design and construction.
- **Accommodates later design decisions**: Since management contracting involves the letting of work in packages, some design decisions, on work that may be sequentially let at the end of the construction phase, may be deferred.

The disadvantages of this procurement method include:

- **Commitment prior to full design**: This is a natural consequence of the overlap of design and construction discussed more fully above.
- **Lack of price certainty**: The client will be uncertain of total project cost until the price of the last works package is known.
- **Increase in client risk**: The management contractor, although in direct contract with the works contractors, is not liable for additional costs arising as a result of their faults, e.g. delays, defective work, claims from other works contractors. This risk lies with the client.

Construction management

This procurement option involves the appointment of a construction manager who is appointed as a consultant and is paid a fee for managing the construction works. Construction management shares a close similarity to management contracting, however, there is a major difference in that, with this method of procurement, the contracts for the works packages are directly with the client. The characteristics of construction management are similar to management contracting. However, because of the direct contract arrangements, additional client involvement is required and this approach is therefore not recommended for those without adequate experience and resources.

Design and manage

This procurement system is similar to management contracting in that a management contractor is paid a fee for managing the construction of the works; however, as the term may suggest, this extends the role to incorporate the management of the design of the project also. The benefits and pitfalls of the system are also similar to those experienced in management contracting.

The major differences relating to the additional design responsibility of the management contractor. To this extent the characteristics of this form of procurement bear a similarity with design and build: the gain of single point responsibility and the loss of design control.

Measurement contracts and cost-reimbursement contracts

Both of these contract types have declined in use in recent years (see section 9.2); however, they provide clients with an option which may be useful in specific situations.

Measurement contracts are used where work can be outlined but not accurately measured. Tenders are submitted on the basis of approximate bills of quantities and work is remeasured when complete and valued in accordance with the prices contained in the approximate bills of quantities. This approach promotes an early construction start but suffers from uncertainty of final price.

Cost-reimbursement contracts are used in emergency situations where construction work is needed to start promptly with little or no design. The contractor will be appointed on the basis of a fee, to allow for overheads and profit, and will be reimbursed the full cost of all materials, labour and plant used during construction.

There are significant disadvantages to this approach. Clients are unlikely to have a meaningful basis upon which to predict final cost, and contractor incentive to complete the works in an efficient manner is missing. Some variants of this method attempt to introduce such an incentive.

9.5 Methods of contractor appointment and payment

There are a variety of alternative approaches to contractor, or works contractor, appointment which may be applicable within each of the procurement options discussed above. These include:

- **Negotiation**: The appointment of the contractor at an early stage of the project on the basis of reputation, competitiveness, convenience, etc. This will improve the time taken to construction commencement, allows pre-construction collaboration and engenders goodwill at the sacrifice of competitive tenders.
- **Two-stage selective tendering**: This, as suggested, involves a first stage whereby contractors submit bids on the basis of approximate documentation and a second stage whereby the contractor submitting the successful stage 1 tender is invited to work with the client to develop the design and contribute to planning matters. This is done on the understanding that a contract will be awarded subject to agreement of price which, where possible, will be based upon the stage 1 tender documentation.
- **Serial contracts**: A series of projects are awarded to a contractor who successfully tenders in competition on the basis of a master bill of quantities which will include a comprehensive range of items and will be used to price each project in the defined series. This (a) achieves the benefits of competition, although some negotiation will be required on each real project, (b) promotes

the development of working relationships and (c) achieves economies through the increased total contract size.

9.6 The need for independent procurement advice

The active selection of an inappropriate form of procurement, or the passive acceptance of a regularly used and hence comfortable practice, can have a major impact on the success of a project.

It is probable that the range of procurement methods will continue to widen, adding further weight to the need for expertise in providing procurement advice. Those involved as consultants within the construction industry are well placed to provide such advice and should ensure that the necessary knowledge and experience are made available to clients.

When giving advice to clients about procurement, advisers should endeavour to give that advice independently and without the intrusion of individual bias. This is true professionalism. It is necessary to acknowledge that there may be a conflict between the objectives of a client and those of a consultant and to ensure that in such situations the client's objectives are foremost. Transparent practices are required on which to build fairness and trust. A client should not be faced with unwittingly selecting an inappropriate form of procurement as a consequence of selecting a particular consultant or profession as the first port of call.

References and bibliography

Clients Construction Forum (1999) *Survey of Construction Clients' Satisfaction 1998–99. Headline Results.*

Construction Industry Board (1997) *Constructing Success: Code of Practice for Clients of the Construction Industry*, Thomas Telford.

Davis, Langdon & Everest (1996) *Contracts in Use: A survey of Building Contracts in Use during 1995*, Royal Institution of Chartered Surveyors.

Franks J. (1998) *Building Procurement Systems: A Client's Guide*, The Chartered Institute of Building and Longman.

Janssens D. E. L. (1991) *Design–Build Explained*, Macmillan Education Ltd.

Latham, Sir M. (1994) *Constructing the Team*, HMSO.

Masterman J. W. E. (1992) *An Introduction to Building Procurement Systems*, E. & F. N. Spon.

Morledge R. and Sharif A. (1996) *The Procurement Guide: A Guide to the Development of an Appropriate Building Procurement Strategy*, Royal Institution of Chartered Surveyors.

Murdoch J. and Hughes W. (1992) *Construction Contracts: Law and Management*, E. & F. N. Spon.

Chapter 10

Risk management

10.1 Introduction

Risk management is not a new concept but a practice that many of us use on a daily basis. As an example, consider briefly our possible concerns and preparations for an overseas vacation. The primary concerns and responses may include:

- *Risk of illness?* We frequently endure vaccinations, commence a course of anti-malarial pills and take out insurance.
- *Risk of loss of belongings?* We may purchase travellers' cheques and/or take out insurance.
- *Risk of theft from home in absence?* We may cancel the milk, advise the neighbours, give items of high value to relatives for safekeeping.

This example demonstrates the importance of active risk management. Some of the principles that we apply instinctively at this personal level are equally relevant to construction; however, due to the complexity and scale of most building projects, good risk management in the construction industry should not rely purely on common sense and instinct.

Risks abound throughout the life of a development project, and include those which may relate to external commercial factors, design, construction and operation. While in this text we have largely considered risk management in the context of the construction phase, the principles involved can, in the main, be transferred to any stage of a project development and may be viewed from several perspectives.

Practitioners in risk management and clients that have adopted its use within their organisations, believe that it brings great benefits. The list of clients using risk management in the UK contains many major organisations, and the promotion of the practice at a high level continues. Risk management has been recognised as a critical factor to the success of projects, in providing the basis for improving value for money in construction. It therefore provides practitioners with an opportunity of contributing further to the added value service they provide.

Table 10.1▸ Risk identification: checklist

Risk category	Indicative examples
Physical	Collapse of sides of trench excavations resulting in delays, additional cost and injury
Disputes	Disruption to a third party's business due to noise or construction traffic resulting in financial loss/litigation
Price	Increased Inflation causing excessive financial loss (easy to ignore in an era of low UK inflation but consider 1970s and overseas locations)
Payment	Delayed payment by main contractor to nominated subcontractors causing programme delays
Supervision	Delays in issuing drawings or instructions resulting in abortive work and claims from the contractor
Materials	Delays in the despatch of unique flooring materials from Italy resulting in programme delay
Labour	Non-availability of labour due to the construction of another nearby major project which causes a regional shortage of specialist subcontractors
Design	Anomalies in the design due to poor communications between engineering consultants and architect resulting in abortive work

10.2 The risk management process

Risk management, in principle, is a simple process in that it requires the assessment of risk and the implementation of a risk management strategy. The assessment of risk first involves *risk identification*, followed by the *analysis of risks* identified. This provides a level of understanding that is necessary to enable the adoption of an appropriate *risk management response*.

Risk identification

The usual method of carrying out a risk analysis is by utilising a workshop at which participants brainstorm suggested risks that they believe could impact upon a project. The workshop forum, which brings together experts from a variety of disciplines, promotes a wide project perspective which, if managed well, will lead to meaningful discussion and communication. This should be considered as a beneficial exercise in itself. In addition to the brainstorming activity, historical data and the use of checklists may also be used, an indicative example of which is shown in Table 10.1.

A generic checklist may be helpful in directing attention to predetermined categories of risk and thus assisting with the identification of those that are project specific. However, there is a danger that their use may restrict thought to those categories contained in the list and it should be borne in mind that this could result in the possible exclusion of major items.

It should be noted that this process is not likely to lead to the identification of all possible risks. The success of the risk identification process will be dependent upon several factors, including the experience and ability of the personnel involved in the workshop, the extent of information readily available, and the skill of the analyst or facilitator.

Table 10.2▸ Risk identification: example scenario

Proposed rail tunnel between Portsmouth and Ryde, Isle of Wight Risk event: explosion during excavation		
Identified sources	Considerations	Possible actions
Inadequate safety measures	Potentially high impact (in terms of time, cost and personal injury)	To reduce the likelihood of this occurrence, ensure that the risk belongs to the contractor
	Implications of statutory health and safety transgression?	Closely examine and monitor health and safety provision.
	Insurance?	Verify insurance provision
Accidental ignition of methane gas	Potentially high impact	Reduce possibility by ground surveys; implement tests and direct work face procedures
	Less opportunity for contractor to control	Closely examine and monitor health and safety provision
	Insurance?	Verify insurance provision
Detonation of an unexploded Second World War device	Potential high impact	Reduce possibility by geo-technical surveys
	Not impossible due to location; infeasible for the contractor to manage?	Consideration of old naval charts/records
	Insurance?	Verify insurance provision

When identifying risk, it is important to appreciate exactly what is needed to be discovered. In order to allow risk analysis and risk management to occur, it is necessary to consider the possible sources of the risk event. For example, if we consider the scenario of a tunnel project, one risk event could be that of an explosion during excavation works. To accommodate risk analysis and risk management, the sources of this risk event should be understood. In this case, these may include inadequate safety measures by workforce, accidental ignition of methane gas or striking a concealed Second World War explosive device. All of these may need to be independently assessed and managed. Table 10.2 provides an indicative example.

Risk analysis

At the outset it is important to be reminded that problems in construction do not confine themselves to cost, although, ultimately, all problems may have an effect on cost. In many situations, time or schedule risk is of more importance than pure cost and, in some cases, quality may be the main priority. Clearly, risk analysis should address the needs of a given situation and should focus upon relevant areas of concern.

There is a range of risk analysis tools which may be used to evaluate the identified risks and the selection of the correct approach will be dependent upon project size, type and opportunity. Examples of some of the approaches, which may be categorised as qualitative, semi-quantitative and quantitative, are discussed below.

Figure 10.1▸ P/I table showing indicative consideration of schedule risk in a tunnel project

Probability	Impact: V. low	Low	Medium	High	V. high
V. low				1	4, 10
Low			6	7	
Medium		2, 3			
High			8		
V. high		5	9		

1. Tunnel collapse
2. Cut through gas mains
3. Cut through power supply
4. Labour dispute
5. Boring equipment breakdown
6. Delay in boring equipment delivery
7. Explosion at boring face
8. Delay due to anti-tunnel protestors
9. Delays due to hard rock
10. Planning delays

10.3 The risk management workshop

Without reference to any particular analysis technique, possibly the simplest and most effective element of risk analysis is the evaluation of risk that is possible during structured workshop discussions. Workshops provide the means whereby risks may be identified, assessed and attended to and great benefit may be obtained without any element of quantitative assessment.

Probability/impact tables (a semi-quantitative approach)

One simple method of evaluating risk is by the use of probability impact tables (P/I tables) which involve the weighting of a qualitative assessment and hence is termed a semi-quantitative analysis. An example of a P/I table – showing an indicative consideration of risks relating to possible schedule delay – is shown in Figure 10.1.

To make probability/impact tables more meaningful it is necessary to define the descriptors, for example, very high probability, equates to more than say, a 75% chance of occurrence. It can be seen that the application of this technique should be designed and applied to meet the circumstances of specific project requirements. The example above, which considers time, may also be adapted and utilised to evaluate risks affecting cost and quality. Use of this simple technique allows risks to be ranked in order of severity and thus allows management to be focused appropriately. There is a need to manage those risks that have a high impact if they arise and have a high chance of occurrence.

10.4 Risk registers

As part of the risk analysis, a database of the identified risks may be produced and used as a management tool. This control document will include key details of each identified risk with possible reference to:

- Description of the risk.
- The predicted probability and impact.
- Ranking of the risk.
- Identification of the owner of the risk.
- Details of the strategy to be adopted to control both impact and probability.
- An action window that identifies the period in the project when the risk may prevail.

The risk register assists as a means of monitoring the management of risks throughout a project and may be used as a reference tool for future project evaluation.

10.5 Simulation (quantitative risk analysis)

It is likely that, on most construction projects, quantitative risk analysis is neither practical nor necessary due to the relative payback from the additional time and expertise required to complete such an appraisal. However, simulation offers a powerful and valuable technique that may be appropriate to large and complex projects.

Construction is an intricate process involving a wide range of activities, each of which may go wrong during the course of a building project. In terms of cost risk, Table 10.3 shows notional costs for a hypothetical project which, for reasons of simplicity, has been reduced to four elements.

The costs shown have been produced following a workshop carried out in the early stages of a project development. Discussions relating to the substructure element may have been recorded in something like the following manner:

> **Substructure**: The quantity surveyor has produced a budget estimate amounting to £170,000. This includes the provision of £20,000 for excavation in bad ground, such as the removal of hard rock. The architect has advised that in a project constructed earlier in the year, on an adjacent site, no rock was encountered. The engineers agree with the architect that rock is unlikely. However, they are concerned that some piling may be

Table 10.3▸ Cost model showing a minimum, maximum and most likely cost

Cost item	Least cost	Most likely	Highest cost
Substructures	150,000	170,000	245,000
External walls	325,000	335,000	345,000
Roof	185,000	195,000	240,000
External works	155,000	215,000	235,000
Totals	815,000	915,000	1,065,000

Figure 10.2▸ Indicative probability distributions of the elements contained in the cost model

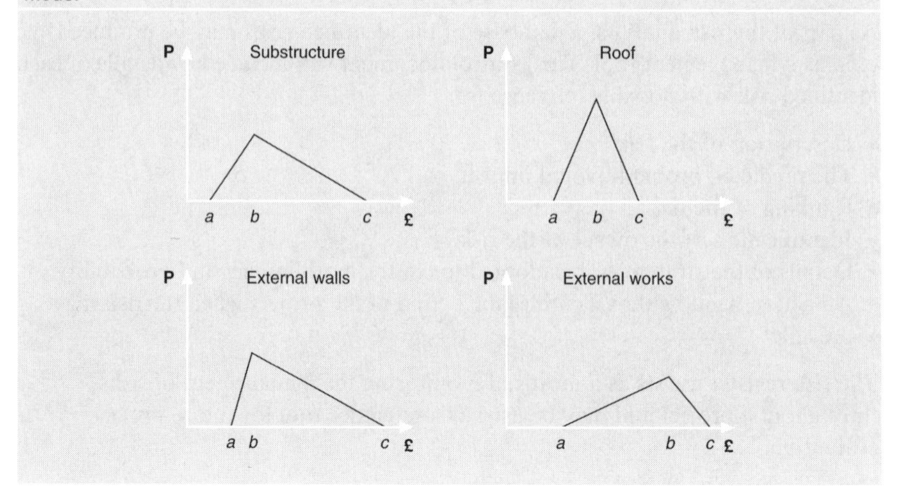

required. No data from site investigations are available at present. The budget prepared by the quantity surveyor is accepted as a reasonable provision for the substructure; however, minimum and maximum costs are also identified.

Similar considerations have been made regarding each of the elements shown in Table 10.3. A question that now needs to be considered is: Does the client have sufficient funds to construct the project? Since this may depend upon the amounts selected, which costs should be used? A pessimistic client or client adviser may select the worst case scenario in each element, resulting in a predicted total cost of £1,065,000. While this approach to risk may be understandable in exceptional circumstances – for example, where a fail safe position is required – it does not hold in most construction projects. In the example used, there may be a 10% chance of the worst case in each element. The chances of each occurring simultaneously is thus $0.1 \times 0.1 \times 0.1 \times 0.1$, with a resultant probability of 0.0001 (i.e. 10,000 : 1). Further consideration of the above cost model will reveal other weaknesses. There is an unmanageable range of 'what-if' scenarios. The values given are discrete, that is 'in between' values are not accommodated; for instance the model does not allow a substructure cost of £160,000. No allowance is made for the fact that the minimum and maximum values are decidedly less probable than the 'most likely' value.

Simulation allows us to model each element in terms of cost possibility. Costs are used in the example, but this can be applied to schedule (programme) risk also. It allows for continuous values, as opposed to discrete values, and accounts for the likelihood of each value by using probability density.

Figure 10.2 shows the triangular probability density distributions representing each of the four elements considered, where probability (P) is on the vertical axis and minimum cost (a), most likely cost (b), and maximum cost (c) are on the horizontal axis.

Figure 10.3▸ An illustration of output from a simulation exercise

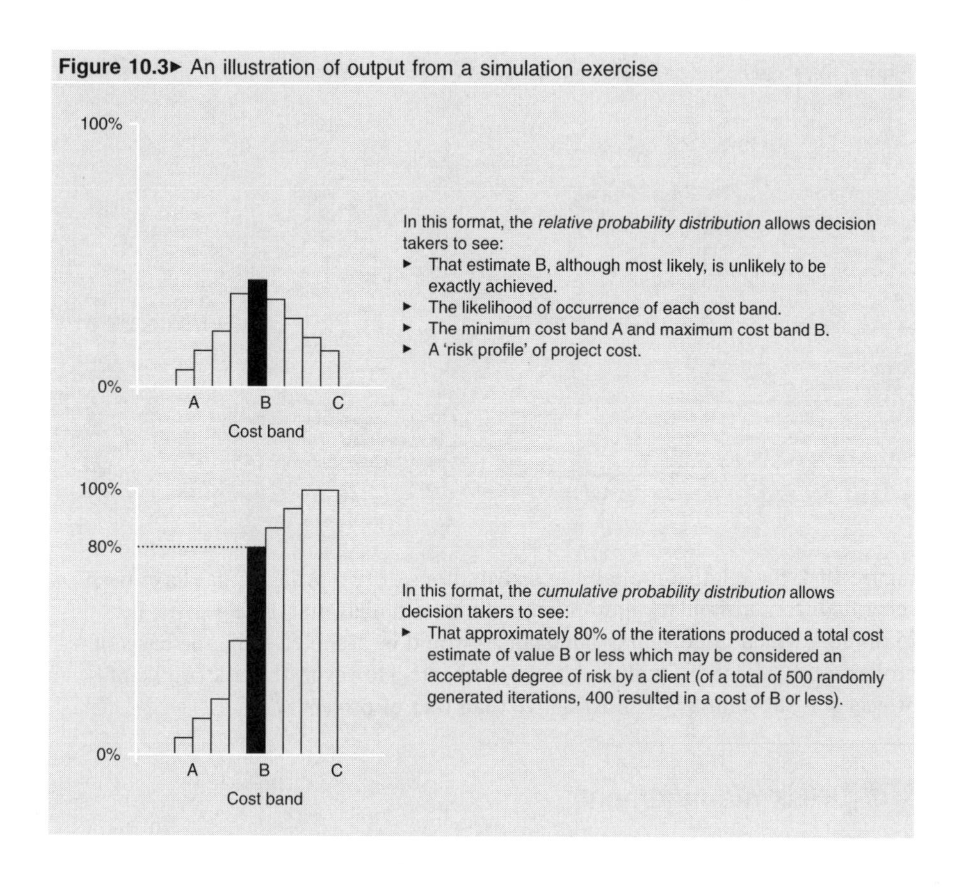

Following the construction of the elemental cost models in the form of probability density distributions, simulation computer software is used to randomly select elemental values which are collected to produce an estimate of total project costs. This exercise or iteration is repeated many times to produce, say, 500 estimates. Since, in each iteration, the selection of values is dependent upon each elemental probability density distribution, most of the elemental values will be selected about point *b*. The frequency at which total project estimates will comprise elemental estimates tending towards values *a* or *c* is very low and the likelihood of an estimate being produced from the sum of four minimum elemental costs (point *a*) or four maximum elemental costs (point *b*) is mathematically unlikely. The output of the simulation exercise can be presented as a relative or cumulative probability distribution. This information provides decision takers with a much clearer picture of the risks involved than by the provision of a single-point estimate.

Illustration of relative and cumulative probability distributions

Information in the form shown in Figure 10.3 allows clients and client advisers to see a full picture of possible project outcomes and therefore assists in decision making. To illustrate the significance of this, consider

Figure 10.4▸ Comparative simulation outputs

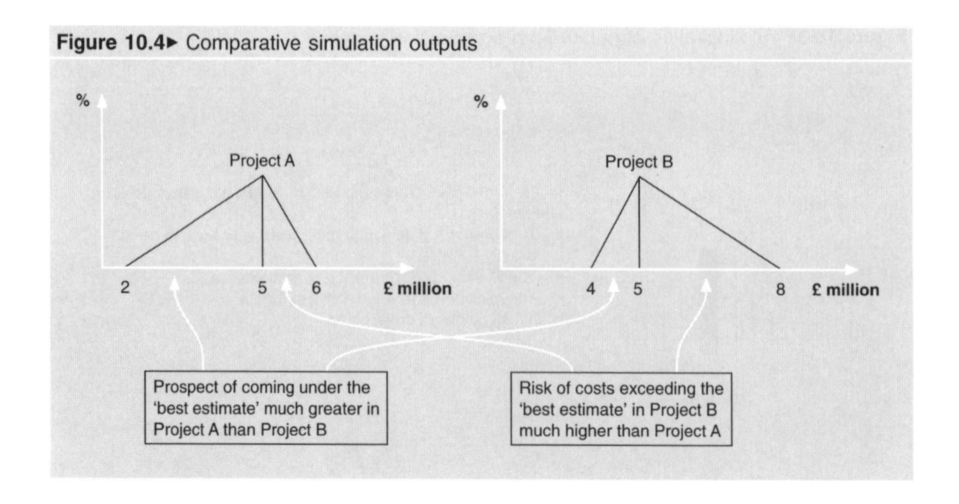

Figure 10.4, the relative probability distributions of two projects that have been generated from a simulation exercise. Both have similar most likely costs, i.e. £5 million, which in normal circumstances would be assumed to be the basis of project estimates traditionally reported to clients. However, the risk profile of project B is substantially less attractive than that of project A.

10.6 Risk management

Once risks have been identified and evaluated, the decision on how to manage them must be made. Effective risk management requires:

- A focus upon the *most significant risks.*
- The consideration of *risk management options.*
- An understanding of effective *risk allocation.*
- An appreciation of the factors which may affect a party's *willingness to accept risk.*
- Consideration of the *response* of a party if and when a risk eventuates.

Focus upon the most significant risks

Methods of determining the most significant risks in a project have been discussed above. While the author has doubts about the need to artificially limit the number of risks to be actively managed, it will be clearly beneficial to concentrate upon those risks which are considered to be high impact/high probability.

Risk management options

Risk management options that are available for consideration may be categorised as follows:

- **Shrink**: A risk can be shrunk or reduced by, for example, establishing more information about a risk situation.

- **Accept**: A risk may be accepted by a party as unavoidable and as being inefficient or impossible to adopt an alternative strategy.
- **Distribute**: A risk may be distributed to another party; for example, a construction client may distribute design risk by selecting a design-and-build contract alternative.
- **Eliminate**: A risk may be eliminated by the abandonment of a project or by the abandonment of a particular element of a project.

This is presented to highlight a convenient, and possibly more memorable to some than others, mnemonic (SADE); however, in so doing, implies no order in the consideration of risk management options. Notwithstanding this, it is sensible to consider the possible reduction of a risk prior to considering further action since the level of the new risk may influence subsequent action.

In addition to the options above, each of which is considered to provide an active response, there are two further passive alternatives:

- Risks may be monitored without action, i.e. keeping an eye on the situation.
- Risks may be unintentionally accepted.

The former of these may be a poor response, and the latter a potential disaster for the unaware recipient!

To demonstrate the effect of the above risk strategy options, consider the following project scenario:

> **Project**: The construction of a block of residential flats in which a double basement car park is required. The flats are positioned in a busy inner city location, surrounded on all sides by main roads/buildings in occupation. The ground is known to contain a significant amount of landfill and existing services (including large Victorian sewers, gas mains and water mains) have been vaguely identified as close to the intended works. Clearly there is an abundance of potentially major risk events in constructing the Basement Car Park! These include: earthworks collapse; cutting through existing services; damage to existing buildings and roads; problems of construction due to confinement of site and bad ground.

Consider the indicative appraisal of risk management options shown in Table 10.4 and considered from the perspective of the client.

It is important to recognise that when risk management action is taken, in each case – including that of the elimination option – *secondary risks* should also be considered. These risks, which are identified in the example below, arise as a result of the selected risk strategy.

There appears to be a view, held by many involved in the construction industry, that the best way to handle risk is to distribute it to another party. We see this throughout the construction hierarchy in Table 10.4: from clients to consultants to contractors to subcontractors. This may be in the belief that such action results in the elimination of the risk, but that, as shown above, is not necessarily the case. It is possible to expand risk by distribution and this should be appreciated. This can be shown by consideration of design and build procurement. The key objective of this method of procurement is single-point responsibility, that includes the transfer of design risk from client, and consultants, to the contractor. Since the design process in design and build is likely to be carried out speculatively

Table 10.4▶ Appraisal of risk management options

Identified risk: earthwork collapse (say high probability, very high impact)		
Risk management option	Possible action	Possible secondary risks
Shrink the risk	Obtain more accurate information about the nature of the site and location of existing services; relocate the car park; direct a specific construction method, e.g. contiguous piling	Additional information may be inaccurate; relocated car parking may produce additional problems; selected construction method may cause problems to contractor resulting in additional problems
Accept the risk	Allow a contingency to cover the eventuality; insure against the risk (insurance in this case is seen as the contribution to a wider contingency fund; some may classify this as distribution)	Contingency provision is inadequate; risk acceptance by client promotes more 'care free' attitude by contractor resulting in more risk eventualities; insurance provision inadequate
Distribute the risk	Pass all associated risks to the main contractor (frequently it would seem, the natural default)	Excessive risk premium; contractor loss resulting in aggressive attitude (e.g. claims) or insolvency; the latter will result in significant problems to the client
Eliminate the risk	Abandon the basement car parking (this may not be a true option without the abandonment of the total project since planners may require the facility)	Location of new car parking causes new problems; piling (bringing new risk) is a requirement of the new substructure design since the poor ground overburden is no longer removed during basement construction

and in a relatively short period, it is reasonable to assume that, at the point of contractual agreement, the design will be less complete than with the traditional procurement option. If it is accepted that the design is less certain, it should also be apparent that the risks relating to the design will be increased. Thus, with design-and-build procurement, design risk is increased, not eliminated! This does not deny the existence of client advantages of this risk transfer but demonstrates the need to fully understand the effects of risk allocation.

Risk allocation

As indicated in the consideration of the active risk management options shown in the above example, risks may be distributed to a third party or accepted by the extant owner of the risk. The effective allocation of risk is a key element of risk management and is a primary function of contracts which should be arranged, from a client's perspective, to optimise the exposure to risk. When considering the allocation of risk to another party, consideration should be given to the following factors:

- The ability of the party to manage the risk.
- The ability of the party to bear the risk if it eventuates.
- The effect the risk allocation will have upon the motivation of the recipient.
- The cost of the risk transfer.

There are many examples of inappropriate risk allocation within the construction industry.

Willingness to accept risk

The willingness with which a party may be prepared to accept a risk will be dependent upon several factors, including:

- **Risk perception**: A party who has recently experienced a serious injury on a construction site is quite likely to perceive the probability of a similar occurrence on a new project more highly than someone without the experience. This viewpoint may be translated into additional risk premium.
- **Risk attitude**: A party who is risk averse is, in essence, someone less willing to accept risk than someone risk seeking. This fundamental will also be translated into the assessment of a risk premium.
- **Ability to bear risk**.
- **Ability to manage risk**.
- **The need to obtain work**: This factor may be the most significant of all. When times are hard, a party is more willing to accept risk as a necessary means of survival. Risk acceptance is therefore market sensitive.

Response when a risk eventuates

If and when things ultimately go wrong, a party who has accepted the corresponding risk will be liable for the related losses. While this is understood, it is also quite likely that the loss sufferer will be motivated towards recovery. In some situations, this may be manifested in contractual claims or a reduction in quality of construction. This reality should be recognised and the mitigation of risk should be recognised as desirable irrespective of who is in ownership.

10.7 The case for risk management

The construction industry has a poor reputation which is due, in the main, to its perceived inability to meet the needs of clients in achieving project completion dates, completing projects within budget and providing a high-quality product. The frequency of the failure of projects to meet the expectations of clients in terms of one or all of these factors is a cause for some concern within the industry. The application of risk management, as briefly outlined above, provides a means of improving this situation. Risk management provides the opportunity to control the occurrence and impact of risk factors and provides clients with better information upon which to make decisions.

Although risk management is being utilised by a growing number of clients, research carried out in 1996 indicated that its use within the domain of the quantity surveyor was very low. The principal method of accommodating risk in most projects being the use of a contingency fund. The methods of determining an appropriate contingency amount appear to be very crude, including reliance upon advice from clients and the use of a standard percentage addition to the estimated contract sum. It would therefore seem that, in many situations, the contingency fund, which is intended to be an all-inclusive risk provision, is ill considered. Since this is currently the only consideration given to risk in many projects, the application of aspects of risk management could be used to greatly improve the accuracy of this risk allowance and would add much value to the service provided to clients.

In several respects, the approach to risk management is similar to that of value management. It usually involves:

- A workshop which is managed by a risk analyst or facilitator.
- Brainstorming and other techniques which assist with the decision-making process.
- A structured approach.
- Multi-discipline participation.

This similarity is acknowledged by some practitioners and organisations who now merge the activities of risk and value management into one inclusive workshop. The use of project workshops dedicated to the consideration of value improvement and risk, not to be confused with the regular activities of design team meetings, appears to offer a good opportunity for adding value to the service we provide to clients.

References and bibliography

Abrahamson M.W. (1984) 'Risk management'. *International Construction Law Review*, pp. 241–264.

Byrne P. (1996) *Risk, Uncertainty and Decision Making in Property Development*, E. & F. N. Spon.

Chapman C. B., Ward S. C. and McDonald M. (1989) *Roles, Responsibilities and Risks in Management Contracts*, SERC Research Grant Report; University of Southampton.

Hogg K. I. and Morledge R. (1995) 'Risks and design and build: keeping a meaningful perspective'. *The Chartered Surveyor Monthly*, May, pp. 32–33.

Institution of Civil Engineers and the Faculty and Institute of Actuaries (1998) *Risk Analysis and Management for Projects* (RAMP), Thomas Telford.

Kelly J. and Male S. (1993) *Value Management in Design and Construction: The Economic Management of Projects*, E. & F. N. Spon.

Latham, Sir M. (1994) *Constructing the Team*, London, HMSO.

Raftery J. (1994) *Risk Analysis in Project Management*, E. & F. N. Spon.

RICS (1995) *Improving Value for Money in Construction: Guidance for Chartered Surveyors and Clients*, University of Reading for the Royal Institution of Chartered Surveyors.

Smith N. J. (ed.) (1999) *Managing Risk in Construction Projects*, Blackwell Science.

Vose D. (1996) *Quantitative Risk Analysis: A Guide to Monte Carlo Simulation Modelling*, John Wiley & Sons.

Chapter 11

Taxation

11.1 Introduction

While taxation is a matter with which we may all be familiar at a personal level, there is a likelihood that, as construction professionals, our understanding of taxation issues is generally deficient. The effects that taxation may have upon construction and, conversely, the effects that construction may have upon tax liability, can have a significant impact upon clients. It is therefore fitting that, since the theme of this book is the provision of added value to construction clients, construction-related taxation and, specifically, the capital allowance system should both be considered.

A difficulty in writing on the subject of taxation is the rapidity and frequency with which the law may change. Readers are alerted to this and are consequently recommended to consider the detail of this chapter as indicative. Finance Acts, the contents of which are introduced at the Budget each year, frequently alter the ground rules. Corporation Tax is subject to annual review, as are the qualifying items, rates of allowance and opportunities for application. Further doubts exist with the progress and intentions of European harmonisation, which is seen as a prerequisite to the free movement of people and capital within the European Union. It is almost certain that, at the time of reading, the rates of taxation and allowances shown (current in April 1999) will have changed. However, the principles relating to the taxation issues are less ephemeral, but by no means stable, and it is this aspect that is considered important.

11.2 Capital allowances

The objectives of the capital allowance system are, in essence, both clear and simple:

- To make an agreed and fair allowance for the depreciating value of a capital investment, and thereby allow businesses to rightfully receive a tax benefit as a valid expense.
- To promote investment in desirable activities and depressed localities.

However, in application the extent of associated litigation indicates that principles of capital allowances may be seen as both hazy and complex.

The system that has been evolving since the late nineteenth century is covered by legislation in the form of the *Capital Allowances Act 1990*. This followed the Capital Allowances Act 1968. Amendments thereto have been incorporated in subsequent Finance Acts issued by Parliament. The law relating to capital allowances will undoubtedly continue to change in response to the objectives of governments and the ability of individuals to continually identify opportunities and develop methods of *avoiding* tax, which is legal, as opposed to *evading* tax, which is not.

To understand the principles of capital allowances, a general knowledge of some key elements of business taxation is required, in particular, the workings of Corporation Tax. An appreciation of the difference between revenue and capital expenditure, and the way in which the depreciation of capital assets is considered by the Inland Revenue, is also required.

Corporation Tax

The tax that a company pays on its profits, which include both income and capital gains, is known as Corporation Tax. This was first introduced in 1965, replacing income tax and profits tax on companies. It is chargeable on companies resident in the UK and on unincorporated associations, e.g. sports clubs. In order to encourage small businesses and new enterprises, there are variable rates of taxation that are dependent upon the size of the company or the level of profits earned. For example, commencing in the financial year 2000, a new starting rate of tax of 10% will apply to profits less than £10,000. The main rates of Corporation Tax that were announced in the 1999 budget, are:

- Normal rate 30%
- Small profits rate on profits up to £300,000, 20%
- Small profits rate on profits £300,000–£1,500,000, marginal relief

In addition, the following rates were announced to be effective in the financial year 2000 (Price Waterhouse Coopers, 1999):

- Small profits rate on profits less than £10,000 10%
- Small profits rate on profits £10,000–£50,000 marginal relief
- Small profits rate on profits £50,000–£300,000 20%

The starting point for the assessment of taxable income is the calculation of net profit, that is revenue (or income) less taxable expenses. Thus, a bicycle manufacturing company, with a revenue of, say, £2,000,000 from the sale of cycles, would deduct revenue expenditure, say £1,500,000, which could include salaries, rent, the purchase of raw materials, etc., to leave a net profit of £500,000. It should be noted that this is not a necessarily taxable profit. In the event that the same company purchased a new machine, e.g. to automatically weld metal tubing, the cost of the equipment would not be considered as revenue expenditure but as an investment, i.e. capital expenditure. This would provide an assessable benefit, not only in the year of purchase but also in future years. The use of the accruals convention dictates that the purchase of assets this year, which will provide an income in the future, may not be claimed as expenses in the year of expenditure. Likewise income received this year for sales from previous years is excluded from this year's tax calculation, this being assessed for taxation purposes at the time of sale. Therefore, in the case of a company purchasing its own premises or items of manufacturing equipment, this means that the element of expenditure relating to income generation in future years, which would normally be the main part, cannot be offset against income in the year of acquiring the fixed asset.

The important point to note is that revenue expenditure is allowable against tax in the year it is incurred while the treatment of capital expenditure is significantly different. Clearly, therefore, it is important to be able to separately identify revenue and capital expenditure. From the discussion above, this would not seem a difficult task. However, there are many situations where a clear line between the two is difficult to draw. For instance, repairs to existing metal cladding will be considered as revenue expenditure, while the cost of its total renewal may become a capital item. The point at which a repair item becomes replacement is in practice often unclear. At the extremes it is generally obvious, but at the margins it may be necessary to test the decision through the courts.

11.3 Capital allowances system

The rationale behind the disparate treatment of capital and revenue expenditure is clear. However, it is equally apparent that since a capital asset does not last for ever, there is a need to allow for the depreciating value of such as a business expense. In the UK, depreciation, which is difficult to assess accurately and consistently, is subject to the interpretation and objectives of accountants. It is also possible to measure it in several different ways. It is thus not considered as a tax-deductible item. In its place, we have a capital allowances system that shows recognition of depreciation by prescribing, by statute, a range of varied allowances against expenditure on fixed assets. If an allowance for depreciation were given, it could result in different companies receiving varied benefits for the same expenditure. This would be inequitable. The capital allowance system reduces this possibility and much of the contention that could arise from the arbitrary assessment of depreciation value. It is therefore effective in simplifying administration and provides more certainty in the prediction of benefit and liability.

Range of allowances

Capital allowances are applicable where, in the course of trade, a business incurs expenditure on specified categories of buildings and on machinery and plant. Table 11.1 summarises the main allowances that are currently available. An explanation of the allowances and examples of their application, relating to industrial buildings and machinery and plant is shown in a later example.

Table 11.1▸ Current writing-down allowances

Qualifying item	Writing-down allowance
Machinery and plant	25% on a reducing balance basis
Industrial building allowance (IBA)	4% on a straight-line basis
Enterprise zones	100% initial allowance
Hotel buildings	4% on a straight-line basis
Agricultural/forestry buildings	4% on a straight-line basis
Scientific research	100% initial allowance
Patents and 'know-how'	25% on a reducing balance basis
Sundry: mines, oil wells, mineral rights, cemeteries, crematoria	4% on a straight-line basis

As with the rates of Corporation Tax, there are additional allowances available that are intended to provide incentives for investment to small and medium sized companies.

It is reasonable to assume from the range of allowances shown in Table 11.1 that the most frequently encountered areas of capital allowance application in practice are in respect of machinery and plant and industrial buildings. Machinery and plant allowances apply to both commercial and industrial buildings and are therefore a consideration in a large number of building projects. Similarly, industrial buildings are widespread. Thus, attention to these particular areas of allowance has been given in some detail throughout.

Machinery and plant

The most complex area of capital allowances concerns machinery and plant, the terms for which are not defined in any tax legislation. Machinery may be defined by reference to the *Oxford English Dictionary*:

- Machinery – *works of a machine.*
- Machine – *apparatus for applying mechanical power, having several parts, each with definite function; bicycle, motorcycle, etc.*

However, the lack of a clear definition for plant has caused and continues to cause difficulties, frequently resulting in litigation, in establishing the validity of possible items of qualifying expenditure (Whitehouse, 1997). The Finance Act 1994 provides some clarification as to what will not be accepted as qualifying plant, but this is of only limited assistance. The case law relating to claims for machinery and plant is extensive and provides guidance as to what may or may not qualify. For an item of plant to be accepted for capital allowance purposes, two broad conditions should be met:

1. *It must be used by the taxpayer in the course of his trade.* The history of relevant case law dates back to *Yarmouth v. France (1887)*, the date of which demonstrates the longevity of litigation relevant to capital allowances and the problems with the definition of the term *plant*. Reference to an extract from the judgement of this case (Lindley LJ), which was not about taxation but a worker's claim for compensation due to injury caused by his employer's plant,

in this case a horse, shows that the use of the item of plant or apparatus is crucial. 'There is no definition of plant in the Act, but in its ordinary sense it includes whatever apparatus is used by a businessman for carrying on his business, not his stock in trade, which he buys or makes for sale: but all goods and chattels fixed or moveable, live or dead, which he keeps for permanent employment within his business'.

It can be clearly seen, therefore, that qualifying plant must be used for the purposes of the business of the claimant and should not be stock in trade. The definition of plant provided by the case was extended in the case of *J. Lyons & Co. Ltd v. AG (1944)* when the distinction was made between plant *with which* a trade is performed and plant which forms part of the setting *in which* the trade is carried out (Saunders, 1998).

2. *It must not be part of the premises or setting in which the business is being carried out.* Although this seems clear, there are situations where an item may be part of the premises or setting and also be a means by which a trade is carried out. This aspect has caused much contention. For example, in *IRC v. Barclay Curle & Co. Ltd (1969)*, an entire dry dock was allowed; while it was the premises of the business, it was also the means by which the trade of the business was performed and allowed by the courts.

In assessing validity, the main consideration to be made is one of functionality. Where doubt exists as to whether or not a claimed item is part of the setting or premises, or of its function, supporting factors may include:

- **Is the item *permanent*?** Permanence in a building may suggest that an item is part of the premises rather than an item of plant.
- **Would the building be *complete* without it?** That the building would be incomplete without the item suggests that it is part of the premises.
- **Does it create an *ambience*?** An item that provides ambience in a restaurant may be accepted as plant since its function is to attract customers (*IRC v. Scottish and Newcastle Breweries (1981)*). Note that such an item could be rejected in an alternative building in which ambience was not considered a function necessary to the trade.
- **Does it provide an *amenity*?** An item that provides an amenity may be accepted as plant since its function is to attract customers (*Cooke v. Beach Station Caravans Ltd (1974)*).

It is important to note that the tests above are not mutually exclusive and provide support to a case rather than certainty. Also, an item of expenditure, which may qualify in one situation, may not in another. This will depend upon the facts of each case and the interpretation of relevant case law, an examination of which suggests that there is often a very fine line between success and failure.

Industrial building allowances

Industrial buildings are defined in detail in the Capital Allowances Act (1990). The buildings that may qualify include: buildings used in a manufacturing process (e.g. mills and factories), buildings used in the storage of raw materials, relevant associated welfare facilities and production offices. Offices as such do not qualify for industrial building allowances, however, the full industrial buildings allowance will apply where an industrial building contains an office unit which is less than 25% of the total cost.

Difficulties may arise where the function of a building is unclear or divided between qualifying and non-qualifying use. For example, a warehouse that is used to store both raw materials (qualifying) and finished manufactured goods (non-qualifying).

The machinery and plant content of industrial buildings will qualify as such and will be assessed at a differing rate of allowance (25% on a reducing balance basis in lieu of 4% on a straight-line basis). An example of this is contained in the scenario below.

Application of allowances

The following example (Tables 11.2–11.5) illustrates in outline the application of capital allowances for a project incorporating both industrial and commercial buildings.

Scenario

A manufacturing client has built a new business centre in the East Midlands. The development incorporates a large factory building and a separate two-storey office building. Assumptions: no capital allowances previously available to the company, development is not in an enterprise zone, client is liable to Corporation Tax at the standard rate of 30%.

Table 11.2► Cost breakdown

Item of expenditure	Cost
Expenditure on industrial buildings	£16,000,000
Expenditure on two-storey offices	£2,000,000
Proportion of machinery and plant allowances:	
Within industrial building (say 25%)	£4,000,000
Within offices (say 35%)	£700,000
Total machinery and plant allowance	£4,700,000
Total IBA (£16,000,000 less £4,000,000)	£12,000,000

Table 11.3► Taxation implications: Year 1

	Without capital allowances	With capital allowances
Profit (subject to allowances), say	£3,000,000	£3,000,000
Industrial building allowance:		
Year 1: £12,000,000 @ 4%	(£0)	(£480,000)
Machinery and plant allowance:		
Year 1: £4,700,000 @ 25%	(£0)	(£1,175,000)
Taxable profit after allowances	£3,000,000	£1,345,000
Corporation tax @ 30%	(£900,000)	(£403,500)
Profit after tax	£2,100,000	£2,596,500

Table 11.4▸ Taxation implications: Year 2 (assumes a 33% increase in profit and no further purchase or disposal of assets)

	Without capital allowances	With capital allowances
Profit (subject to allowances), say	£4,000,000	£4,000,000
Industrial building allowance:		
Year 1: £12,000,000 @ 4%	(£0)	(£480,000)
Machinery and plant allowance:		
Year 1: £3,525,000 @ 25%	(£0)	(£881,250)
Taxable profit after allowances	£4,000,000	£2,638,750
Corporation tax @ 30%	(£1,200,000)	(£791,625)
Profit after tax	£2,800,000	£3,208,375

Table 11.5▸ Taxation implications: Year 3 (assumes a 10% increase in profit, the purchase of additional machinery and plant valued at £500,000 and no disposal of assets)

	Without capital allowances	With capital allowances
Profit (subject to allowances), say	£4,400,000	£4,400,000
Industrial building allowance:		
Year 1: £12,000,000 @ 4%	(£0)	(£480,000)
Machinery and plant allowance:		
Year 1: £3,143,750 @ 25%	(£0)	(£785,938)
Taxable profit after allowances	£4,400,000	£3,134,062
Corporation tax @ 30%	(£1,320,000)	(£940,219)
Profit after tax	£3,080,000	£3,459,781

Notes

- The calculations for Years 1 and 2 illustrate the difference between straight-line and reducing balance write down. Industrial building allowance for Year 2, calculation as Year 1. Machinery and plant calculation shows balance reduced from Year 1 to Year 2 (i.e. from £4,700,000 to £3,525,000).
- In practice, machinery and plant expenditure is claimed for via the use of a pool of allowable items. The calculation for the value of the plant and machinery allowance shown in the annual assessment for Year 3 could be as follows:

Value of the pool brought forward from Year 2:	£2,643,750
Value of new assets in Year 3:	£500,000
Total value of machinery and plant allowance for Year 3:	£3,143,750

- The outline includes an assessment without the application of capital allowances for comparison purposes.
- The impact of capital allowances is demonstrated by this indicative example; likewise the impact of failing to make a claim.

11.4 Enterprise zones

As a means of encouraging investment in areas considered as deprived, some developments in designated enterprise zones – i.e. designated by the Secretary of

State – may qualify for tax relief. These can include commercial projects, and may qualify for 100% of the total cost as an initial allowance. This incentive will clearly assist in directing some investment towards areas of high unemployment.

11.5 Purchased property

In addition to the availability of capital allowances for a tax payer who has incurred capital expenditure on the construction of owner-occupied buildings, allowances are also available for investment in existing property. 'All purchased properties, irrespective of age, will contain an element of machinery and plant in differing degrees' (Blakeley, 1992). The data shown in Table 11.6 have been prepared from an analysis of approximately 100 capital allowance claims relating to purchased buildings.

The opportunity to reduce tax liability through the application of capital allowances relating to purchased property is clear. This may result in the improvement of yields on property investment, which is demonstrated in Table 11.7. Note that, in Table 11.7:

- Plant allowances = 25% writing-down allowance per annum on a reducing balance.
- Corporation Tax rate = 30%.

Table 11.6▸ Indication of the proportional values of machinery and plant for a range of building types

Building type	Machinery and plant proportion
Computer centres	25–37.5%
Hotels: luxury	20–26%
Hotels: normal	13–21.5%
Offices: air-conditioned	18–24%
Offices: modern but not air-conditioned	12.5–20%
Offices: old	8–13%
Industrial	5–10%
Shop shells	2.5–5%
High-tech development	7.5–12%

Source: Blakeley (1992)

Table 11.7▸ Reducing tax liabilities

Year	Reducing balance	Plant allowances	Cash benefit generated	Yield	
				Net CA	Including CA
1	£4,800,000	£1,200,000	£360,000	6.30%	8.10%
2	£3,600,000	£900,000	£270,000	6.30%	7.65%
3	£2,700,000	£675,000	£202,500	6.30%	7.31%
4	£2,025,000	£506,250	£151,875	6.30%	7.06%
5	£1,518,750	£379,688	£113,906	6.30%	6.87%

Source: adapted from Whittaker (1995); adjusted to reflect revised rates of Corporation Tax

- Value of machinery and plant = £4,800,000.
- Total tax saved over time = £1,440,000 (£4,800,000 @ 30%).
- Assumed property value = £20,000,000; assumed yield net of capital allowances, 9% less 30% Corporation Tax.
- Yield incorporating benefit of capital allowances e.g. Year 1 = ((20,000,000 × 0.063) + £360,000) / £20,000,000.

11.6 Partnerships and sole traders

Partnerships and sole traders are not liable to Corporation Tax. Individuals pay income tax on profits or share of profits earned. However, they are eligible to claim for capital allowances. Therefore, the principles of the system described above are also applicable to these groups, but with varying rates of taxation.

11.7 Quantity surveying involvement

The potential large impact that capital allowances may have upon company profits after tax has been indicated by the example above. Quantity surveyors should be aware of the scope afforded to clients by the capital allowance system, in respect of both new build and property acquisition, and appreciate the significance of their possible role in providing advice on both opportunity and application.

Market opportunity

The provision of advice on capital allowances has the added complexity of falling between two distinct professions. Taxation is within the professional domain of the accountant. However, construction expertise, which is also required, lies elsewhere, most likely in the domain of quantity surveyors. The apparent gap in the necessary consolidated expertise has given rise to the emergence of capital allowance experts specialising in property and construction. Several major quantity surveying practices offer this specialist service and some large firms of accountants now employ quantity surveyors to provide the otherwise missing construction knowledge base. This will undoubtedly benefit some organisations; however, evidence indicates that many clients are failing to obtain their capital allowance entitlement (Blakeley, 1992). Some of these may be small and occasional clients who may not receive the level of advice offered to larger corporations. Opportunity therefore exists for the involvement of construction professionals who may identify capital allowance potential and provide all clients with advice that will make their businesses more tax efficient. The fees relating to the provision of this advisory service are also tax deductible.

Advice may be provided to clients, client's accountants or directly to the Inland Revenue. It should be borne in mind that the quantity surveyor adviser is unlikely to be able to overview a client's business operations or overall tax position and is therefore limited by this restriction. It should also be borne in mind that the provision of advice relating to taxation might not be adequately contained within the protective cover of existing professional indemnity insurance. This needs to be clarified before venturing into these activities.

Assessment of machinery and plant value

The possible range and complexity of machinery and plant items which may be incorporated in many projects requires and provides scope for involvement by quantity surveyors who may assist in their identification and evaluation. In addition to the value of the claimed items, an assessment of the value of associated builder's work, preliminaries, fees, relevant variations and claims must also be calculated. This distinction, for taxation purposes, between plant and buildings is often complex, misunderstood and may need to be interpreted by the courts and relevant case laws. These aspects are generally also beyond those of the knowledge of accountants. The exact allocation of some of these additional costs to single items of machinery and plant may be difficult or impossible to ascertain and approximations may be required. The co-operation of the contractor and subcontractors will probably be necessary in many situations to provide additional cost information and supporting documentation. For example, a breakdown of a lump sum for electrical installations is necessary to determine the separate cost of the power supply relating to a specified qualifying item of equipment such as an air-conditioning unit. This will qualify, whereas the power supply generally will not. Design-and-build projects may provide an additional difficulty in that cost data will probably be more difficult to obtain.

Tax planning

Some careful pre-planning to accommodate future claims for capital allowances may be possible and could assist, for example, in the identification and separation of relevant claimed items (and associated works) in contract documentation. Also, design considerations could incorporate those relating to capital allowances, e.g. the incorporation of non-industrial elements within rather than apart from an industrial building unit to extend possible industrial building allowances.

Ethical considerations

This chapter ends with a note relating to ethics. Research carried out by Morledge and Hogg (1997) found compelling evidence of unethical practice relating to the area of financial management and accounting within the construction industry. While it is possible that, to some, this is an accepted part of the industry which is mitigated by common knowledge within the construction fraternity, similar attitudes may not exist within the Inland Revenue. Attempts to over-value a possible claim for plant and machinery in order to enhance a tax benefit may be regarded as an attempt to evade rather than avoid taxation. The penalties for this may be severe. It is not illegal to avoid the payment of tax, and skill and judgement can be used to enhance claims for capital allowances without crossing this ethical and legal boundary. Many years ago a House of Lords judgement stated: 'No man in this country is under the smallest obligation, moral or otherwise, to arrange his affairs as to enable the Inland Revenue to put its largest possible shovel into his stores.'

References and bibliography

Blakeley P. (1992); Value for money. *Taxation*, 11 June, pp. 262–265.

Magrin E. (1993) Plant and machinery identifier. *Taxation*, 18 November, pp. 146–150.

Morledge R. and Hogg K. I. (1997) The perceptions of ethics in the practice. In B. Greenhalgh (ed.) *Practice Management for Land, Construction and Property Professionals*, E. & F. N. Spon.

Nellis H. G. and Parker D. (1992) *The Essence of Business Taxation*, Prentice Hall.

Price Waterhouse Coopers (1999) *Budget Analysis 1999*, Price Waterhouse Coopers.

Saunders G. (ed.) (1998) *Tolley's Taxation Planning (1998–99)*, Tolley Publishing Company Ltd.

The Chartered Institute of Taxation (1999) *Corporation Taxation*, 10 May. Available at http://www.taxation.org.uk/html/facts/corp.htm.

Whitehouse C. (1997) *Revenue Law, Principles and Practice*, Butterworth.

Whittaker R. (1995) Digging deep for hidden taxation savings. *The Chartered Surveyor Monthly*, March, pp. 44–45.

Chapter 12

Change and innovation

12.1 The changing nature of the construction industry

The pattern of investment in the construction industry in the United Kingdom during the past thirty years has been characterised by many different factors. While there have been fluctuations in its workload, these have tended to follow the previous patterns of boom and bust. At its peak, turnover has been almost as high as £58 billion and approximately 6% of the gross domestic product (GDP). It directly employs about 1.4 million people. In 1971, 52% of all construction work was in the public sector. Due largely to privatisation, this had declined to less than 20% by the late 1990s. Its relative proportion increased during the recession years of the early 1990s, but this slipped back again towards the end of the millennium. During this period there has also been some shift away from new build towards repair and maintenance. New build was over 65% in 1970 and now represents less than 50%.

The number of construction firms has also boomed during this same period. The increase in number from 70,000 in 1970 to almost 200,000 by the turn of the twenty-first century is partly reflected by the last statistic, but is due largely to the way in which the industry has become organised. Subcontracting has always been a feature of the industry but now it has become a way of life. Coupled with this scenario, the number of large firms (those with in excess of 1200 employees) have declined over the same period of time from almost 80 to less than 30 (Harvey and Ashworth, 1997). The construction firm in the UK appears to be following the same path as the motor car industry – fewer in number and more foreign owned. The dramatic asset swap which took place between Wimpey and Tarmac in 1996 is a recent high-profile example within the construction industry.

There have also been changes in the way in which work is procured. There has been an increasing trend towards single-point responsibility, illustrated by the increase in design and build and its counterparts. The Private Finance Initiative (PFI), instigated by government to build public sector projects with private capital, has been a partial success, although the somewhat bureaucratic process has disinterested some contractors.

The fierce competition for work has reduced tender margins, as illustrated by the comparisons produced by the Building Cost Information Service between building costs and tender prices. Profitability in the mid-1990s resulted in around 1% average pre-tax profits on turnover and 2% on the capital employed. This was despite interest rates on borrowings of 2–4% over a base rate of 6%.

Clients too have become more vocal in their criticism of the industry, citing poor quality, long-time building and what they believe to be too high prices. The former grouping of the British Property Federation in the 1970s were perhaps the first to suggest that buildings and the process of their construction could be done better. One response to this has been to place an emphasis on the management of the design and construction process. This has frequently resulted in a further tier of administration with mixed degrees of success. Some processes have been borrowed from manufacturing industry, at times forgetting that the construction site bears little resemblance to the well-ordered production lines now operated by extensive robotic systems in manufacturing industry.

A forward-looking review of Britain's construction industry, *Building Britain 2001* (Centre for Strategic Studies in Construction, 1988) provided a broadly based and provocative vision of the construction industry for the turn of the twenty-first century. However, some of its predictions have now been overtaken by other events.

Why is change taking place?

- Government intervention in the construction industry through privatisation, PFI (Private Finance Initiative), CDM (Construction Design and Management) Regulations, compulsory competitive tendering, European legislation.
- Recent reports on the state of the industry:
  - *Constructing the Team* (Latham, 1994).
  - *Improving Value for Money in Construction* (Atkin and Flanagan, 1995).
  - *Towards a 30% Productivity Improvement in Construction* (Construction Industry Board, 1996).
  - *A Statement on the Construction Industry* (Barlow, 1996).
  - *Value for Money: Helping the UK Afford the Buildings it Likes* (Gray, 1996).
  - *The Challenge of Change* (Powell, 1998).
  - *Rethinking Construction* (Egan, 1998).
- Pressure groups formed to encourage change and improvement.
- International comparisons, particularly the USA and Japan, and the Single European market.
- The apparent failure of the construction industry to satisfy the perceived needs of its customers, particularly in the way that it organises and executes projects.
- The influence of patterns of education and research on these processes.
- Trends generally in society towards greater efficiency, effectiveness and economy.
- Rapid changes expected from information technology in respect of design, management and manufacturing processes and practices.
- A desire to make the construction industry more high-technology oriented.
- The varying attitudes among the professions.

- Developments occurring in other similar and different industries and the need for the construction industry to catch up.
- A desire to reduce the adversarial nature associated with the construction of buildings.
- The awareness of Quality Assurance mechanisms in other industries.
- A desire to establish best practices in construction work.
- The over-riding wish of clients for single-point responsibility.
- Changes in culture and work practices.

The construction industry is perceived as dirty, dangerous, exposed to bad weather, unhealthy, insecure, underpaid, of low status and poor career prospects for educated people (Latham, 1994). It is widely agreed within the industry that it is too easy to set up in business as a general contractor. No qualifications, no experience and virtually no capital are required. While market forces ultimately remove incompetent firms by depriving them of work, the existence of such unskilled producers is bad for clients and damages the wider reputation of the industry. Women are seriously under-represented at all levels in the industry (Latham, 1994).

12.2 The challenge of change

The British construction industry has a long and honourable tradition and records of achievement, both within the UK and overseas. A few years ago the same could be said of the British car manufacturing industry. While some notable parts of the latter have survived, many have ceased to exist or have declined at an alarming rate. Those that have managed to survive made radical changes to their practices. There are already sufficient signs to indicate that the construction industry of Great Britain is facing similar challenges (Barlow, 1996).

The changes required first are mainly organisational and cultural rather than of a technical nature. While increased resources are necessary in information technology and research and development, the underlying cultural changes remain of paramount importance.

The issues facing the future of the construction industry are many and varied and include:

- Under-capitalisation exacerbated by fragmentation and the large numbers of small firms.
- Low technology, labour intensive and traditionally craft based.
- Litigious basis for settling differences and disputes.
- Lack of prototype development resulting in untested and ill-specified components and technologies.
- Low level of information technology.
- The use by government of the construction industry as an economic regulator contributes towards a cyclical workload.
- High number of company business failures.
- Poor image, working practices and employment conditions.
- Difficulties of recruiting, training and retaining a skilled and committed workforce.

There is a lack of understanding of best practices or even an awareness in some situations that change is the order of the day and urgent. There is a determination to survive but a lack of realisation of international competition.

Experience suggests that while improvement among the leaders in the construction industry can be expected to match the best in the world, the improvement generally will take time and will involve radical changes in culture and probably its structure and organisation. Barlow (1996) has suggested that the following are examples of some of the best practices that can be learned from the manufacturing industry sector:

Focus on customer satisfaction

Recognise that clients want buildings and support after completion, at the right price, appropriate quality and standards, on time and meeting their needs.

Attention to the process as well as the product

Product design has now become a byword in manufacturing industry, where the processes used have contributed towards increasing the appropriateness of the product. Research is necessary in the construction industry in process analysis.

Concept of total quality approach and attitudes

The total quality approach should not be confused with total quality management or quality assurance, which are now widely accepted in the industry. Total quality programmes are often expensive to implement, relying on extensive training to bring about a shift in culture. It represents a continuous improvement programme.

Benchmarking

The practice of benchmarking all of a company's activities against the best competition or against organisations who are known to be industry leaders is now commonplace in some quarters (see Chapter 8). A characteristic of many of these companies is their willingness to share knowledge with others.

Team-working and partnering, including supply chains

This aims to make every individual feel worth while and, as such, it leads to greater pride within a company. It aims to harness the intelligence and experience of the whole workforce. It also extends beyond the individual company to include consultants, contractors, subcontractors and suppliers. For large clients they are also often part of the team (see Chapter 7).

Information technology

The construction industry must welcome the more widespread use of information technologies and embrace the current Technology Foresight Initiative.

Patel (1999) has argued that for a company to succeed it must change to meet new demands. Companies that fail to respond to change will surely diminish in their importance and in the worst cases will disappear from the scene. Patel argues that a number of leading companies have already shown how this can be done. He cites a number of companies who barely existed even a few years ago. These include:

- *Orange* developed through ideas and innovation.
- *Microsoft* that has harnessed information technology in a new way.
- *Intel* who has focused on the importance of the value chain.
- *Goldfish* who has changed the rules of the UK credit card market.
- *Cisco* has become a world-wide leader in providing networking solutions.
- *Oracle* who has quickly identified the pivotal role of customers.

12.3 Innovation

Innovation is about the introduction of the new in place of the old, especially in changes of customary practices. Innovation involves developing a strategy which involves people, practices, processes and technology to deliver high added value.

A great deal of technological change passes unnoticed. It consists of the small-scale progressive modification of products and processes. Such has been the description of the construction industry. Freeman (1987) has described such changes as incremental innovations. They are important, but their effects or shock-waves are only felt within the immediate vicinity. More important are the radical innovations or discontinuous events which can have a drastic effect upon products and processes. A single radical innovation will not have a widespread effect on the economic system. Its economic impact remains relatively small and localised unless a whole cluster of radical events are linked together in the rise of new industries or services, such as the semiconductor business. These are the more significant changes. The following five generic technologies have created new technology systems:

- Information technology
- Biotechnology
- Materials technology
- Energy technology
- Space technology

These represent new technology systems that change the style of production and management throughout the system. The introduction of the electronic computer is an example of such deep-going transformations.

The concept of long-waves developments, each of less than fifty years' duration, is generally associated with the work of the Russian economist, N. D. Kondratiev. Figure 12.1 is a highly simplified picture of the sequence that might be commonly envisaged. Four complete K-waves are identified with the implication that we are currently entering a fifth. Each wave has lasted approximately fifty years and appears to be subdivided into the four phases of: prosperity, recession, depression and recovery. Each wave also seems to be associated with significant technological innovation associated with production, distribution and organisation.

The fifth Kondratriev cycle, which appears to have begun in the early 1990s, is associated primarily with the first of the five generic technologies identified in Figure 12.1. The next wave of technological and economic changes will cluster around information technology (Freeman, 1987). The development of information

Figure 12.1► Kondratiev long waves of economic activity and associated major technologies

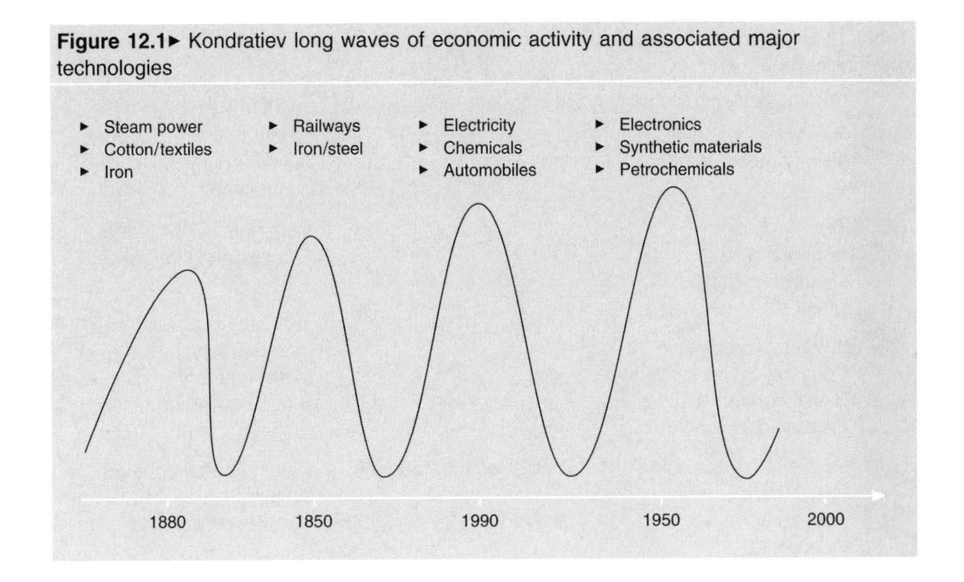

technology originates from communications technology and computer technology, as shown in Table 12.1.

Innovation in the construction industry should seek to:

- Identify important innovations, which have contributed to construction quality cost effectiveness and added value.
- Educate industry leadership in the importance of innovation within their respective companies and associations.
- Encourage architects, engineers, surveyors and contractors to develop and implement processes for innovation to occur.
- Recognise important innovations within the industry, once achieved, through awards, publications, etc.
- Develop financial support for further innovations within the construction industry and its varied services.

The Construction Productivity Network (CPN) was formed as a result of the Latham (1994) report and is under the auspices of the Construction Industry Board. This is a partnership of the construction industry, its clients and government working together to improve efficiency and effectiveness in construction. The Board aims to secure a culture of co-operation, team work and continuous improvement in the industry's performance. CPN exists to promote the sharing of knowledge and the benefits of innovation across all sectors of the construction industry through workshops, visits, conferences and publications.

12.4 Information technology

Information technology continues to develop at an exponential rate. Those involved in the construction industry now have extensive access to this technology.

Table 12.1▸ Information technology: the convergence of technologies of computer and communications

	Communications technology	Information technologies	Computer technologies
1940	Radio Military mobile radio	–	Single function computers General purpose computers
1950	Tape recording Cable television Microwave links Direct distance calling	–	Commercial computers Programming languages
1960	Video tape recording Communication satellites Digital communications Electronic switching	–	Integrated circuits Minicomputers Structured programming
1970	Facsimile transmission Mobile radio	Professional databases	Database management systems
	Teletext Videodisks	On-line enquiry	Microprocessors
1980	Teleconferencing	Management information systems	Spreadsheets
		Computer-aided design	Portable computers
	Local area networks	Integrated text and data storage	Optical disk storage
	Cellular radio	Computer-aided manufacture Computer-aided diagnostics	Expert systems
	Wide area networks	Materials planning and scheduling	Voice recognition
	Personal telephones	Electronic mail	
1990		Teleconferencing	Parallel processing Learning capability
	Mobile satellite communications	Remote-sensing devices	Natural language recognition Optical chips
2000			Biochips Ultra-intelligent machines

Source: adapted from Freeman (1987)

What can be imagined will be achievable, if this is desired. Many aspects currently not imagined will also be achieved, probably in the short term. There is a tendency to overestimate what will happen in the next few years, but to underestimate what may take place in the medium term.

Information technology has been shown to be an effective tool for a wide range of applications in the construction industry. These have included computer-aided design (CAD), drafting and assisting manufacturers of building materials. By the use of information technology, design professionals have been able to demonstrate considerable success in the modelling of design solutions. Although information technology applications are capable of achieving high work levels and have been reported to offer time savings of up to 40%, they have failed to meet the expectations for increased productivity and product quality in the construction industry. The opportunities are substantial. The following areas lend themselves to the development of information technology in construction applications:

- Design and production techniques, which incorporate design aids, robots, energy management, commissioning of buildings and education and training.
- Information systems, which employ databases, quantities, drawings and models, specifications, property data and electronic data interchange.
- Hardware and software, which include interfaces, expert systems, standards, integration of applications and software techniques.
- Communications, which apply to intelligent buildings, wide area networks, local area networks, integrated services digital networks, optical fibres and wiring, radio technology and security.

Despite the opportunities for the greater use of information technology in the construction industry, its application falls far behind the opportunities that are available. There remains a lack of appropriate hardware and software, although this has improved considerably over the past ten years in terms of ease of use and reliability. There remains a lack of understanding about computers by the senior management, and this is sometimes seen as a boast. The high technology revolution of the latter half of the twentieth century has transformed manufacturing processes and applications throughout almost every sector of British industry and, moreover, has created a whole new sector of industry. A glaring exception to this is the British construction industry, which operates largely with low technology, low skill and is labour intensive (Powell, 1995).

A survey by Barbour Index in 1997 on the use of information technology in the construction industry suggested the following:

- All professions are using computer-aided design.
- 83% of professionals have access to a portable computer.
- 50% have a CD-ROM on the machine.
- 25% have access to the Internet.
- 50% expect their expenditure on IT to increase.

Overall, construction continues to spend less on information technology than almost any other industry. For example, the financial services sector invests 5% of its annual turnover, manufacturing sets aside 2%, but construction only 0.5%. Such statistics have contributed, in part, to the suggestion that the construction industry is backward, its technological growth is retarded, and it exists largely as a nineteenth-century handicrafts technology.

12.5 Rethinking construction

Rethinking Construction (Egan, 1998) recognises that construction in the UK at its best is excellent. It has developed engineering ingenuity and a design flair that is recognised throughout the world. Its capability to deliver the most difficult and innovative projects matches that of any other construction industry in the world. The report recognises that the industry is, as a whole, under-achieving. It has low profitability and invests too little in research, development and training. Also, too many clients are dissatisfied with its overall performance.

Five key areas for change were identified in the report:

- **Committed leadership**: The need for vision and change at all levels.
- **Focus on the client**: The view is that the customer is always right.
- **Integration of processes and teams**: Fragmentation of operations affects success.
- **Quality driven agenda**: About getting it right first time.
- **Commitment to people**: Providing appropriate health and safety, wages and site conditions.

The report also stated that training and quality are inextricably linked. Real reductions in construction costs, while at the same time maintaining value, are unlikely to reduce significantly until the education of its workforce is improved through a culture of team work. The report suggested that employers, the Government and the National Training Organisations (NTOs) should put together an agenda for urgent action.

Performance indicators

It is easy to suggest that improvements in processes and practices are being achieved based on a subjective judgement and anecdotal evidence alone. There is also often a resistance to want to measure or attempt to quantify such changes. It is also all too easy to distort the data, unless clear and precise guidelines are employed. In some cases in the past, improvements have occurred and their effect has then been attributed to a particular cause. Upon further investigation the cause and effect are not linked. For example, the 30% reduction in cost identified by Latham (1994) may appear to be achieved largely due to the suppressed costs – both labour and materials – of the recession in the mid-1990s. It may be difficult, when looking back to the start of the twenty-first century, to identify whether improved methods of working actually achieved this goal, or whether natural effects of improved technologies were the real reasons. The debates that raged throughout the 1980s about the poor time performance of UK building when compared with countries abroad, were only partly remedied through productivity agreements. Other improvements in time performance were often restricted because of the different regulations and organisation that was adopted in the UK.

In this context it is important that the construction industry sets itself clear and measurable objectives. These might be achieved through the use of performance indicators or quantified targets. Measures of improvement will be required in terms of cost, time and quality, relevant to the aims and objectives of the individual client. The targets must be real and composite. They must not be achieved through cutting corners in other respects, such as safety and wages. In order to make such gains last, and thereby add value, continuous improvement must be implemented.

Egan (1998) identified a number of measures designed for sustained improvement. These are shown in Table 12.2.

Table 12.2▸ Performance indicators

Indicator	Improvement per year	Definition
Capital cost	Reduce by 10%	All costs excluding land and finance
Construction time	Reduce by 10%	Time from client approval to practical completion
Predictability	Increase by 20%	Number of projects completed within time and budget
Defects	Reduce by 20%	Reduction in the number of defects at hand-over
Accidents	Reduce by 20%	Reduction in the number of reportable accidents
Productivity	Increase by 10%	Increase in value added per head
Turnover	Increase by 10%	Turnover of construction firms
Profits	Increase by 10%	Profits of construction firms

Source: Egan (1998)

12.6 Competitive advantage

Porter (1980) describes a key contribution for shaping and reshaping thinking in the current context of world economies. In competitive advantage the rationale is not directed towards organisational structures or change, but with profitability as the strategic driver. Porter argues that there are five competitive forces that determine profitability:

- The potential of new entrants into the industry.
- The bargaining power of customers.
- The threat of substitute products.
- The bargaining powers of suppliers.
- The activities of existing competitors.

One of the most important concepts established by Porter is that of the *value chain*. This is a systematic way of examining all the activities a firm performs and how these interact. Primary activities are inbound logistics, outbound logistics, marketing, sales and service. Support activities include procurement, technology development, human resource management and infrastructure. The way in which one activity in the chain interacts with another can be crucial. This can occur within the organisation, or externally with suppliers.

Porter argues that a firm gains competitive advantage by performing these activities, alone or linked, more economically or in a better way than its competitors.

12.7 Business process re-engineering

The importance of this technique is to learn as much as possible from the success and failure of other industries who have had to respond to massive cultural changes. Improvement, or added value, can be achieved through construction re-engineering. No other concept has recently received more interest and criticism than re-engineering, because it is a concept that is easy to understand but difficult to put into practice. Hammer (1990) suggested that successful re-engineering projects are founded on six basic principles:

- Organise around outcomes not tasks.
- Have those who produce the output of the process perform the process.
- Subsume information processing into the real work that produces the information.
- Treat geographically dispersed resources as though they were centralised.
- Link parallel activities instead of integrating their tasks.
- Put the decision point where work is performed, and build control into the process.

In spite of a few well-known examples of success, there is much evidence to suggest that re-engineering fails or, at best, produces only marginal results in a majority of organisations in which it is implemented. In some cases this is due to the fact that the programmes are not sufficiently radical and only tinker with the most easily accessible processes.

12.8 Quality assurance

The quality movement has found its most influential expression in total quality management (TQM). This is a philosophical approach to organisational development that constantly focuses the efforts of every employee on maintaining and improving an organisation's services or products. The father of the quality movement is believed to be W. Edwards Denning (1900–1993), who while working at the US National Bureau of the Census observed that the techniques employed for ensuring quality control, while adequate for correcting defects, were inadequate for improving the way things were done. Unless the processes were changed, mistakes would continue to be repeated. The core of Denning's philosophy was that in order to improve the quality of production over time, the people doing the job have to be involved in the improvement of the production process. Denning further argued that the employees' intellect and sense of responsibility have to participate in the quality process as well as their physical effort. The key component in this exercise of winning worker commitment consists of the team sharing a vision of continuous improvement. Denning's philosophy is succinctly summarised as follows:

- The abolition of:
  - Constricting performance appraisals.
  - Performance-related pay.

  - Productivity quotas.
  - Bonus payments.

- The introduction of:
  - Continuous training on the job.
  - Education and self-improvement programmes.
  - Breaking down barriers between departments.
  - Driving out fear in order that people can work effectively.
  - Removing barriers that robbed people of their right to pride of workmanship.

While Denning has been criticised for his approach to performance appraisal and his dismissal of management by objectives, he is credited with an enormous contribution to the thinking behind the post-war economic miracle in Japan. The focus of this has been towards continuous quality improvement.

12.9 Lean construction

This process, which has been derived from Japan, is concerned with the elimination of waste activities and processes that create no added value. Lean production is the generic version of the Toyota Production System, recognised as the most efficient production system in the world today. Incidentally, Toyota's activities in the construction industry are larger than those of its car manufacturing industry. It is acknowledged that construction is quite different from the automobile industry in which spectacular advances in productivity and quality have been achieved in the past ten years. Construction, by comparison, remains the most fragmented of all industries, and Egan (1998) saw that as both a strength and a weakness. The waste, or *muda* as it is termed in Japan, can include:

- Mistakes or reworking.
- Bad programming.
- Products and services that do not meet client's needs.
- Not conforming to the specification.

Lean thinking has been borrowed from manufacturing industry processes and is aimed at delivering what clients want, on time and with zero defects. Lean construction has identified poor design information that results in a large amount of redesign work as an area for correction. Several organisations around the world have established themselves as centres for lean construction development. The aim, for example, of the *Lean Construction Institute* in the USA is a dedication towards eliminating waste and increasing value.

As very few products or services are provided by only one organisation, the elimination of waste has to be pursued throughout the whole value stream, including those who make any contribution to the process. Removing wasted effort, therefore, represents the biggest opportunity for performance improvement.

Several companies around the world are attempting to introduce lean construction methods into their core businesses. Egan (1998), for example,

Figure 12.2▸ Different contractual arrangements and their effects on time

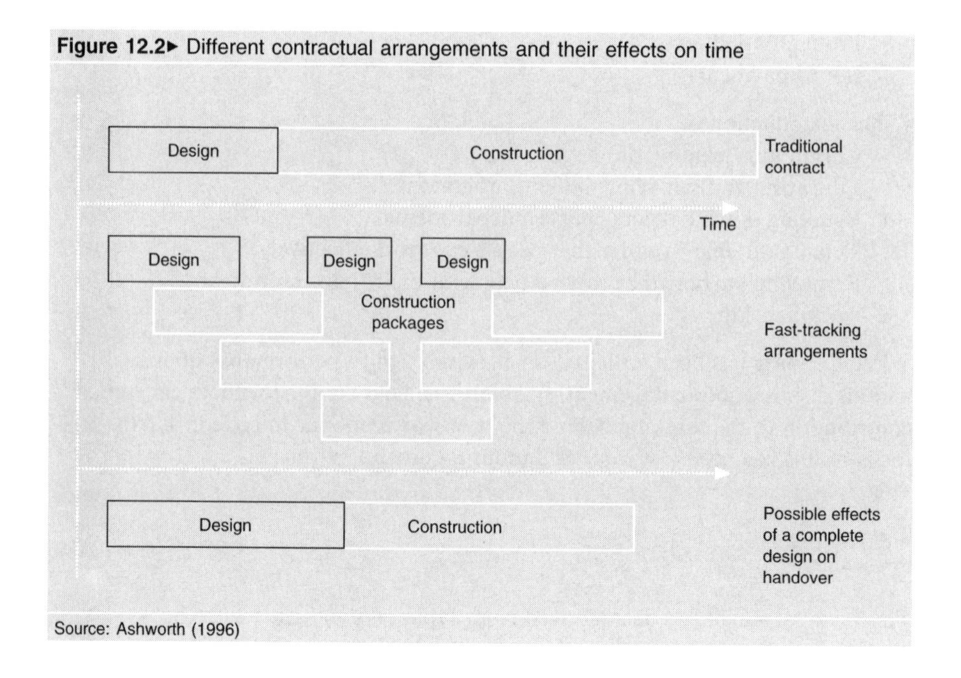

Source: Ashworth (1996)

refers to two firms, one based in Colorado and another in San Francisco. One of these firms has already reduced project times and costs by 30% through developments such as:

- Improving the flow of work on site.
- Using dedicated design teams.
- Innovation in design and assembly.
- Supporting subcontractors in developing tools for improving processes.

This suggests that perhaps the most useful way of achieving cost reductions while still maintaining value is to consider profitable ways of reducing the time spent on construction work on site. The principle of design readiness is the same approach. This suggests that a fully completed design, prior to starting work on site, will save construction time and the respective costs that are involved (see Figure 12.2).

12.10 Conclusions

Added value in design and construction is not solely about reducing building costs, although this has an important part to play. It is more about developing a cultural change in all areas of the industry. The industry is, however, beginning to make that big shift. Integrated teams are being formed in the industry's cultural shift. The industry is beginning to get tough on waste and the causes of waste. The first steps in the new construction culture will be in the direction of adding value and measuring performance through the medium of key performance indicators.

The change being experienced is likely to be ongoing. Quality improvements must be continuous, at least for the foreseeable future.

The Department of the Environment's Construction Sponsorship Directorate (1995) is providing a new focus for change through its research and technology support programme. In this it is seeking partnerships within the industry to develop a whole industry research strategy. Representatives will come together in the Construction Research and Innovation Strategy Panel (CRISP). The strategy seeks to build on existing industry improvement initiatives.

Patel (1999) has identified ten key questions to ask regarding change and adding value in business:

- What are the key discontinuities facing our customers and industry?
- How are the boundaries being blurred in our industry?
- How do I see the future?
- Which particular future do we wish to create?
- What is our ambition or dream and what I am prepared to do about this?
- What innovation do I want to see happen in our industry?
- What change would be relevant and add value?
- What assumptions do I need to challenge about ourselves, our customers, our competitors and our markets?
- What are the forces around that can help us to add value?
- What might be the implications for our company and our people?

References and bibliography

Ashworth A. (1996) *Contractual Procedures in the Construction Industry*, Longman.

Atkin B. and Flanagan R. (1995) *Improving Value for Money in Construction*, Royal Institution of Chartered Surveyors.

Barlow J. (1996) *A Statement on the Construction Industry*, The Royal Academy of Engineering.

Barbour Index (1997) *Use of Information Technology in the Construction Industry*, Royal Institute of British Architects.

Centre for Strategic Studies in Construction (1988) *Building Britain 2001*, University of Reading.

Construction Industry Board (1996) *Towards a 30% Productivity Improvement in Construction*, Thomas Telford.

Construction Sponsorship Directorate (1995) *An Introduction to the Whole Industry Research Strategy*, Department of the Environment.

Egan J. (1998) *Rethinking Construction*, Department of the Environment, Transport and the Regions. HMSO.

Freeman C. (1987) The challenge of new technologies. In *Interdependence and Co-operation in Tomorrow's World*, OECD, Paris.

Gray C. (1996) *Value for Money: Helping the UK afford the Buildings it Likes*, Reading Construction Forum.

Hammer M. (1990) Re-engineering work: don't automate, obliterate. *Harvard Business Review*, July–August.

Harvey R. C. and Ashworth A. (1997) *The Construction Industry of Great Britain*, Butterworth-Heinemann.

Latham M. (1994) *Constructing the Team*, HMSO.
Patel K. (1999) *Change the Game, Change the Rules of the Game*, KPMG.
Porter M. (1980) *Competitive Strategy: Techniques for Analysing Industries and Competitors*, John Wiley.
Powell C. (1998) *The Challenge of Change*, Royal Institution of Chartered Surveyors.
Powell J. (1995) *Towards a New Construction Culture*, Chartered Institute of Building.

Index